Activity Workbook with Real-World Applications for College Algebra

Revised Printing

Dr. Gisela Acosta-Carrasquillo
Valencia Community College

Prof. Margie Fernandez-Karwowski
Valencia Community College

Revised Printing 2010

ISBN 978-0-7575-7364-4

Printed in the United States of America
10 9 8 7 6

To our families for their continuous love and support in all our endeavors.
To our students in hopes that they see how algebra is used in real-world situations.

Contents

Preface

This workbook is designed to supplement your college algebra course. We have provided activities modeling real-world situations that can be used for class discussion, review of topics, or discovery of concepts. Throughout the activities, students will solve problems using multiple approaches: verbally, numerically, graphically, and symbolically. The workbook contains worksheets designed for both independent and group work.

Students will have the opportunity to learn by following step-by-step guidelines and apply the graphing calculator in problem-solving, thus integrating this technology as a learning tool. We have included an appendix that explains use of a graphing calculator and keystrokes for various topics frequently addressed in a college algebra course.

We hope that instructors and students enjoy these activities as they provide opportunities to foster reasoning and problem-solving skills and to experience how college algebra ties into everyday life.

Dr. Gisela Acosta-Carrasquillo　　　　Prof. Margie Fernandez-Karwowski

I. Linear Functions and Mathematical Modeling

Functions: Definition

1. Determine whether each of the following represents a function. Explain why or why not. (Hint: Is the second variable a function of the first variable?)

a.

x	0	4	7	24	50
y	9	12	-6	58	-19

Circle one: Function Not a function

Explain: ____

b.

Sport	*Medals earned*
Baseball	5
Bowling	9
Cycling	7
Gymnastics	9
Hockey	4
Tennis	7

Circle one: Function Not a function

Explain: ____

c. A person; DNA

Circle one: Function Not a function

Explain: ____

d. Letters; mailboxes

Circle one: Function Not a function

Explain: ____

e. Children's heights; children's weights

Circle one: Function Not a function

Explain: ____

f.

Correct answers on a quiz	4	15	7	9	4	18
Score	8	25	16	19	10	30

Circle one: Function Not a function

Explain: ______________________________

2. Determine whether the graphs below represent functions and justify your answer.

a. 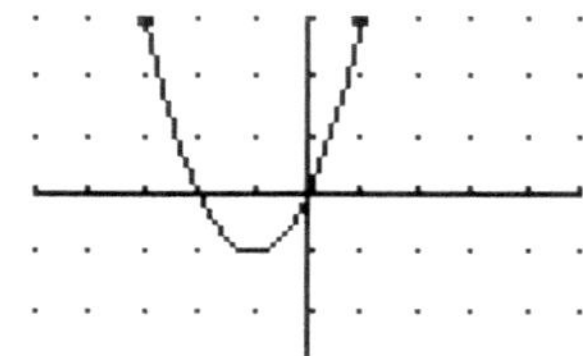

Circle one: Function Not a function

Explain: ______________________________

b. 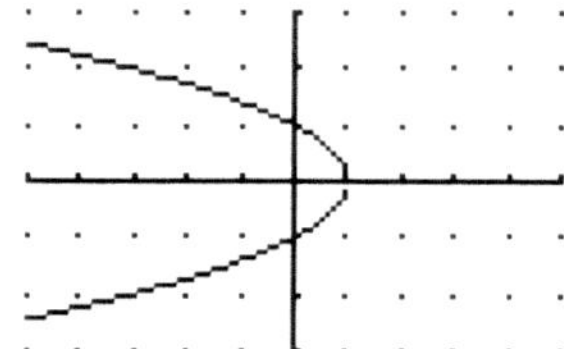

Circle one: Function Not a function

Explain: ______________________________

c. 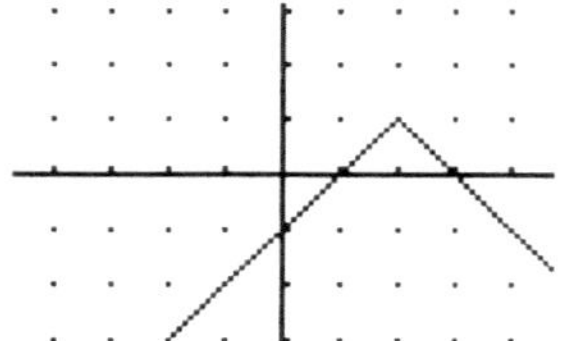

Circle one: Function Not a function

Explain: ______________________________

d. 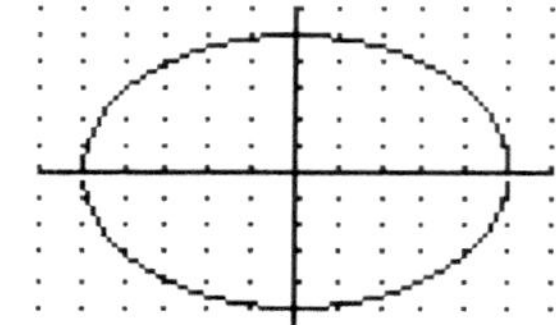

Circle one: Function Not a function

Explain: ______________________________

Functions: Notation and Evaluation

1. The following table denotes the Pepsi 400 winners' average speed (in miles per hour) in Daytona Beach, Florida, where $x = 1$ corresponds to the year 2001.

x (year)	1	2	3	4	5	6	7
r (average speed)	157.601 Dale Earnhardt, Jr.	135.952 Michael Waltrip	166.109 Greg Biffle	145.117 Jeff Gordon	131.016 Tony Stewart	153.143 Tony Stewart	138.983 Jamie McMurray

Source: http://sportsnetwork.com

a. Is r a function of x? Explain. ____________________

b. In which year did the winner have the fastest average speed? __________

c. From the table, find $r(7)$. ____________________

d. Interpret your result from part (c). ____________________

2. An employee of a sales company receives a salary of $26,000 plus 2.5% of the number of sales. The function $S(x) = 26{,}000 + 0.025x$ describes the employee's salary, where x represents the number of sales.

a. Complete the table of values:

X	20,000	80,000	100,000	800,000	1,000,000
$S(x)$					

b. Evaluate $S(950{,}000)$ and explain what it means.

c. Suppose $S(x) = 60{,}000$. Find the value of x, and interpret your answer.

3. The following table approximates the number of tornadoes, T, reported in the United States during each month in 2006, where $x = 1$ corresponds to January and $x = 12$ corresponds to December.

x (month)	1	2	3	4	5	6	7	8	9	10	11	12
T (no. of tornadoes)	47	12	150	245	139	120	71	80	84	76	42	40

Source: http://www.spc.noaa.gov

a. How many tornadoes touched down in August? ______________________

b. Is T a function of x? Explain. ______________________

c. Which month had 76 reported tornadoes? ______________________

d. Which month had the most? ______________________

e. From the table above, find $T(5)$ ______________________

f. Interpret your answer from part (e). ______________________

4. Given $f(x) = 7.9x^4 - |x| + x^2$ complete the table by finding the following:

a. $f(-2)$

b. $f(17)$

c. $f(3.6)$

x			
$f(x)$			

5. Complete the table below for the following function.

$y = 2.6x^3 - 3x^2 + \sqrt{x}$

x	12	15	18
y			

6. Use the graph below to answer the questions:

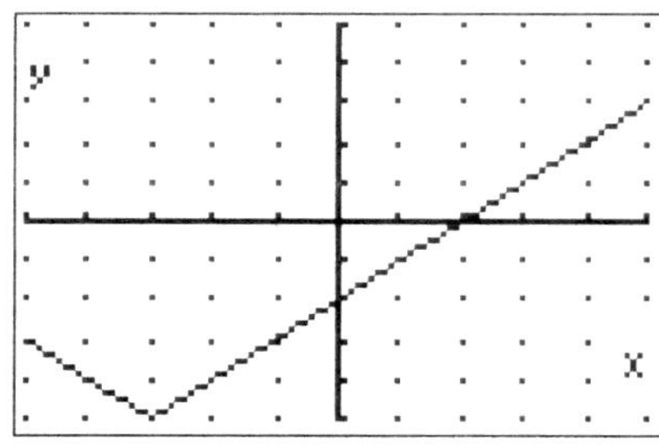

The x and y scales = 1.

a. Find $f(4)$.
b. Find $f(-1)$.
c. Find x so that $f(x) = -4$.
d. Find $f(0)$.
e. What do the coordinates found on (d) represent?
f. Find x so that $f(x) = 0$.
g. What do the coordinates found on (f) represent?
h. How do we know this graph represents a function?

Answers:

a. ______________________________

b. ______________________________

c. ______________________________

d. ______________________________

e. ______________________________

f. ______________________________

g. ______________________________

h. ______________________________

Slope

1. Determine the slope for the following graphs. The x and y scales = 1.

a. Slope ___________  b. Slope ___________ c. Slope___________

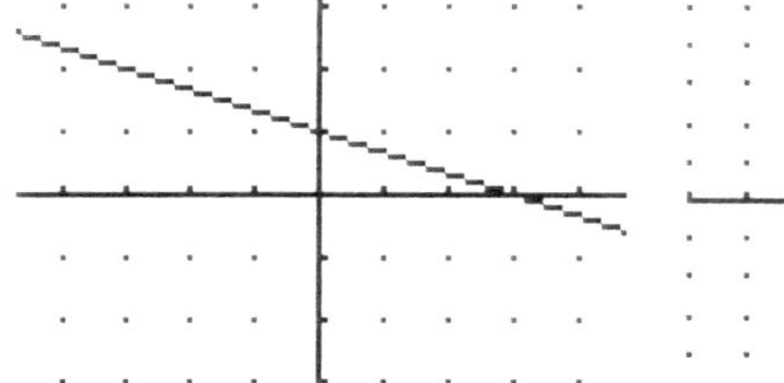

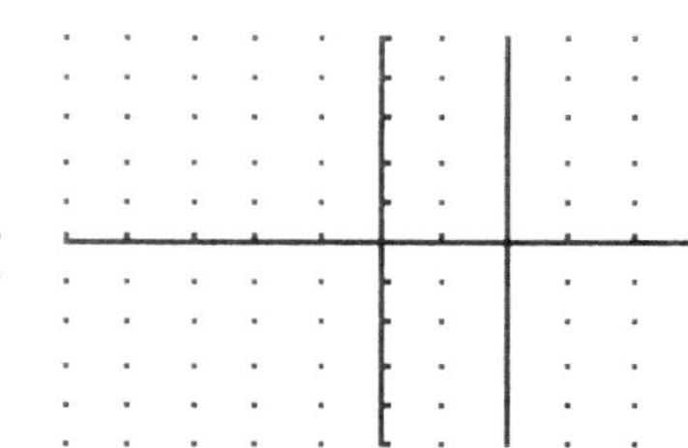

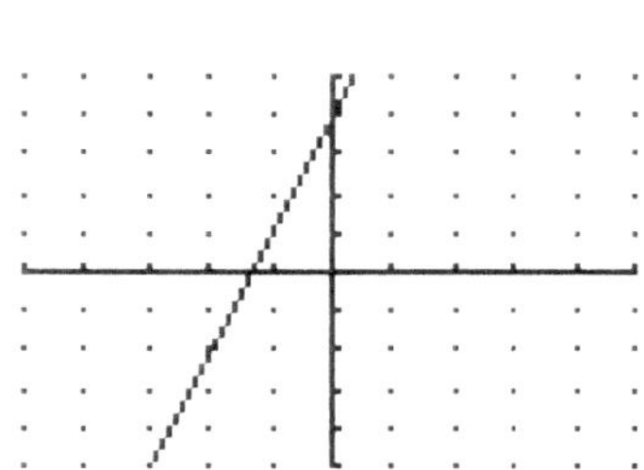

2. Using the "rise over run" concept, draw a line that has the given slope.

a. m = - 4 b. m = 3/2 c. m = 0

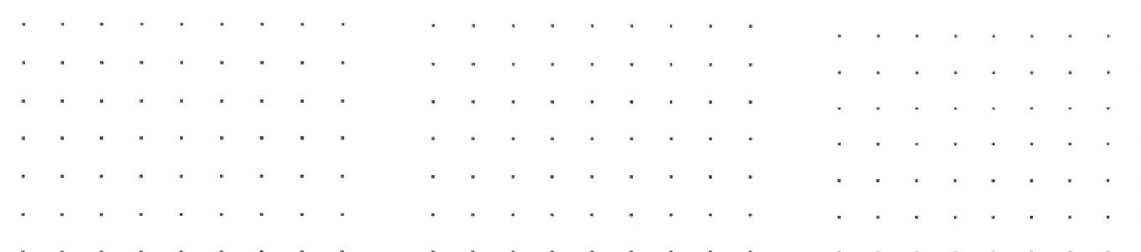

3. Does the following table represent variables that are related by a linear equation? Why or why not? (Hint: Use the concept of slope.)

X	Y
4	-8
7	-5
22	10
38	42

Circle one: Linear Nonlinear

Explain: ____

Slope as Rate of Change and Vertical Intercept

1. A custom golf shirts company charges a design fee of \$15 and sells each shirt for \$35. The total cost, y, can be represented by the equation $y = 35x + 15$, where x represents the number of golf shirts.

a. What is the slope (rate of change) of the total cost equation?

b. State the significance of the slope in the context of this problem.

c. What is the vertical intercept (y-intercept) of the total cost equation?

d. State the significance of the vertical intercept in the context of this problem.

2. According to health experts, the average male body frame should be 106 pounds for the first 5 feet of height plus 6 pounds for each additional inch.
Source: http://www.nlm.nih.gov

a. Determine the vertical intercept and explain its meaning in the context of this problem.

b. Determine the slope and explain its meaning in the context of this problem.

3. According to the automobile club AAA, on Tuesday, February 27, 2008, regular gasoline was selling at a nationwide average of \$3.14 a gallon, up from \$2.35 a year ago.
Source: http://www.iht.com

a. Letting x be the number of years since February 27, 2007, find two data points.

b. Find the slope between the two data points and use appropriate units.

c. Explain the meaning of the slope in the context of this problem.

4. Vladimir and some friends decided to go canoeing during the weekend. They paid a deposit of $45 to rent a canoe and $5.75 per hour.

a. Write an equation in which y represents the total cost of renting the canoe and x represents the number of hours they canoed.

b. What is the rate of change (slope) of the total cost equation?

c. State the significance of the slope in the context of this problem.

d. Find the vertical intercept of the total cost equation and explain its meaning in the context of this problem.

5. The distance in miles covered by a marathon cyclist is given by $d = 15t$, where t is the number of hours the competitor rides.
Source: http://saltlakecitymarathon.com

a. Make a table of values for the distance equation for the following times in minutes: 15, 30, 45, 60, 75, and 90. (Hint: Since t is the number of hours riding, remember to change minutes to the equivalent time in hours.)

t	D

b. Compute the rate of change.

c. What is the significance of the slope in the context of this problem? Answer in a complete sentence.

d. State the vertical intercept and explain its meaning in the context of this problem. Answer in a complete sentence.

Equations of Lines and Their Graphs

1. Lisa owns several vacation beach homes in Florida. The off-season rental rate for each vacation home is \$375 per day, for a maximum of six people using three bedrooms. There is an additional one-time payment of \$250 that includes a cleaning fee and the charge for utilities.

a. Write a verbal equation that expresses the cost of renting one of Lisa's vacation beach homes.

Cost = ________________ + ________________

b. Using C for the total cost and t for time in days, write an equation for C in terms of t.

c. Complete the table of values for the cost of renting a vacation beach home for the following lengths of time.

t	0	3	5	6	10
C					

d. Using graph paper, plot the points and construct a graph. (Choose appropriate scales and label the axes in the context of this problem.)

e. Use your equation to determine the cost of renting a vacation beach home for 14 days. Answer in a complete sentence.

f. If a family of six paid Lisa \$2,875 for the total rental cost, for how many days did they rent the vacation home? Answer in a complete sentence.

2. Ana and Biagio opened a small business selling boxes of gourmet chocolate cookies. They invested \$4,700 to start their business. They sell each box of chocolate cookies for \$30.

a. Write an equation that states Ana and Biagio's profit, P, in terms of the number of boxes of gourmet chocolate cookies, n, sold.

b. How many boxes would they need to sell to make a profit of \$10,000?

3. An appliance repairwoman charges \$50 for diagnosis and \$15 per hour she works.

a. Write an equation for the total cost, C, of repairing an appliance that takes h hours of work.

b. Graph the equation with your graphing calculator. Be sure to use an appropriate viewing window. Include a rough sketch on separate graph paper.

c. Find the slope and vertical intercept of the total cost equation.

d. State the significance of the slope and the vertical intercept in the context of this problem.

e. Calculate the number of hours worked if the total cost for repairing an appliance is \$102.50. Locate your answer on your graph from part (b) and compare both values.

4. Drinking alcohol and driving dramatically increases the probability of being in a crash. According to Florida Department of Motor Vehicles (DMV) records, there were 34,638 DUI (driving under the influence) convictions in Florida in 2006. For drivers over age 21, the legal limit in Florida is a blood alcohol level of 0.08; that is, a breath-alcohol level of 0.08 or higher is prima facie evidence that a driver is under the influence of alcohol to the extent that his/her normal faculties are impaired.

The following table shows the number of drinks consumed and corresponding blood alcohol level for a 175-pound man.

Number of drinks	Blood alcohol level
4	.100
5	.125
6	.150
7	.175
8	.200

Sources: http://www.dmvflorida.org; http://brown.edu

a. Find the rate of change (slope) of the blood alcohol level with respect to the number of drinks. Is this a constant value?

b. Use the grid below to construct a graph for this data, using the number of drinks as your horizontal axis and the blood alcohol level as your vertical axis.

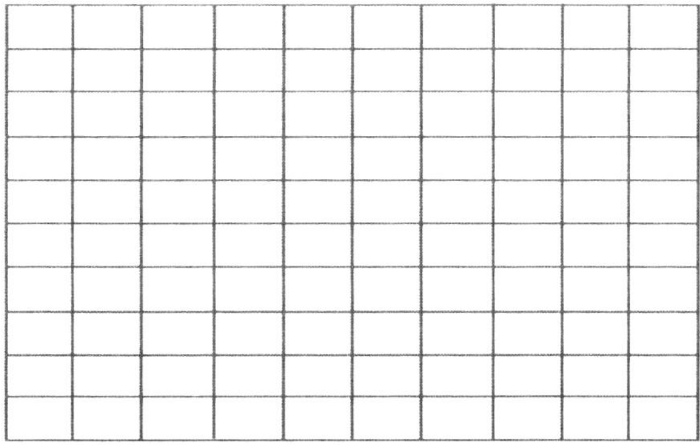

c. Write an equation for this linear model using the point-slope form of a line. (Hint: Use the rate of change and choose one data point from the table.)

d. Use your equation to determine how many drinks a 175-pound man would be able to consume without achieving Florida's legal definition of intoxication. Approximate your answer to the nearest integer.

5. After a long afternoon of studying, Gerard decides to go to McDonald's. While he is eating, he researches on his laptop and realizes that his Big Mac has 540 calories, the medium fries order has 380, and his 16-ounce Coke contains 150 calories. When he gets home, his sister Liana, who works at a doctor's office, asks him to join her on a bike ride. While on the bike ride, she informs Gerard that they are burning 292 calories per hour.

a. What was Gerard's total calorie intake at McDonald's?

b. Write an equation for the total calories burned, C, as a function of hours, h, that he is exercising while biking with his sister.

c. Calculate the number of calories Gerard burns in 1.2 hours.

d. How long will he have to bike to burn off all the calories he ate?
Consider that Gerald started with a total number of calories from part (a) and that he is burning off a number of calories per hour from part (b).

Below is a chart of some of the menu items at McDonald's. Choose one of your favorite items from each column; feel free to select fries or not, and don't forget to order a drink.

Cheese burger	Quarter Pounder with cheese	Big Mac	Filet fish	Grilled chicken	Snack wrap	3-piece Chicken select	6-piece Nuggets	Southwest salad with grilled chicken	Southwest dressing
300	510	540	380	420	330	380	250	320	100

Small fries	Medium fries	Large fries
250	380	570

Source: http://mcdonalds.com

16 oz. Coke	16 oz. Diet Coke	16 oz. Sprite	16 oz. Hi-C Orange
150	0	150	160

e. Calculate your total number of calories.

The next chart lists the average calories burned per hour for each activity. The table is assuming a person of weight 170 pounds. Choose one of your favorite activities from the following chart.

Aerobics low-impact	Basketball	Bicycling < 10 mph	Bowling	Jogging @ 5 mph	Softball/ baseball	Tae kwon do	Tennis singles	Walking @ 2 mph
365	584	292	219	584	365	739	584	183

Source: http://mayoclinic.com

f. Write an equation in terms of the number of calories burned, C, as a function of the hours exercised, h, for your activity.

g. Use your answers from parts (e) and (f) to help you determine how long it will take to burn off your McDonald's calories.

h. Compare your time in part (g) with one of your classmates' time.

Lines of Best Fit/Linear Regression

1. Theresa wants to take hang gliding lessons during her vacation. The lessons cost \$590 for a 4-hour course and \$1,470 for a 12-hour course. Both prices include a fixed initial fee.

a. The variables for this problem are cost and hours. Which variable corresponds to the independent variable, and which to the dependent variable, and *why*?

b. Find the coordinates of two points for a linear model that would fit this data and compute the slope of the line through these points.

c. Use the point-slope formula to determine the equation relating the variables, and write your equation in slope-intercept form.

d. What is the meaning of the slope in the context of this problem?

e. Find the vertical intercept of this linear model and state its significance in the context of this problem.

f. Graph the equation with your graphing calculator. Be sure to use an appropriate viewing window. Include a rough sketch on separate graph paper.

g. If Theresa wants to take a 15-hour course, how much would it cost? Locate your answer on your graph from part (f) and compare both values.

2. The scatterplot in the following figure shows the U.S. median age at first marriage for women, where x represents the number of years since 1940.
Source: http://www.census.gov

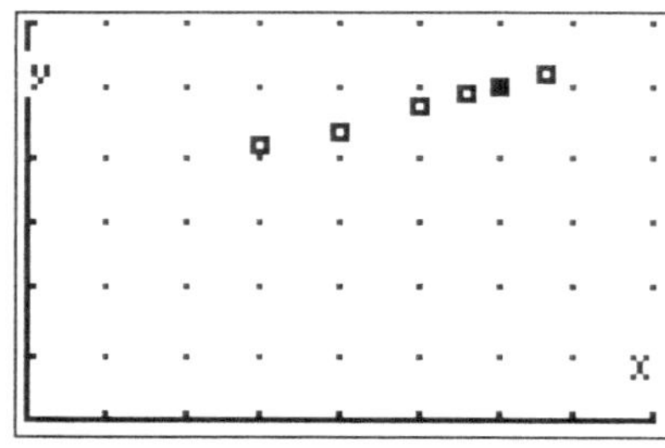

X min = 0, X max = 80, X scale = 10
Y min = 0, Y max = 30, Y scale = 5

a. Using a straightedge, draw a line by hand that best fits the data points.

b. Use your line (regression line) to estimate the U.S. median age at first marriage for women for the years 1970 and 1985.

c. Use the data points from part (b) to approximate the equation of the regression line.

d. Use your equation to estimate the median age at first marriage for women in the year 2005. How does this compare with the data displayed on your regression line from part (a)?

e. Use your equation for the regression line to predict the median age at first marriage for women in the year 2015. Is the answer reasonable?

Take it further with your grapher!

f. The points on the scatterplot in the figure for problem (2) are: (30, 20.8), (40, 22.0), (50, 23.9), (56, 24.8), (60, 25.1), and (66, 25.9). Use your graphing calculator to find the least squares regression line.

g. Use the regression line from your calculator to predict the median age at first marriage for women in the year 2015. How does this compare with your answer to problem (e)?

3. The table below shows the number of households in the United States from 1940 to 2000.

Year	Number of Households (in millions)
1940	34.9
1950	43.5
1960	52.6
1970	63.5
1980	80.4
1990	91.9
2000	103.2

Source: http://www.census.gov

a. Let t represent the number of years since 1930 and use the following grid to construct a scatterplot of this data. Label your axes and identify your scales.

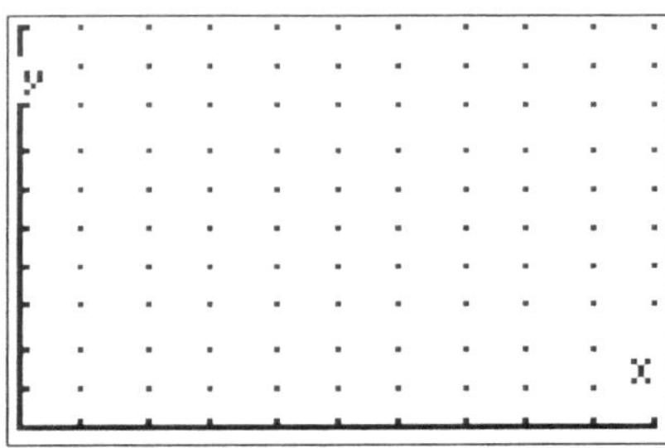

b. Draw a line of best fit (regression line) for the data points.

c. Choose two points and find the equation for your regression line.

d. Use your linear model to estimate the number of households in 1995 and compare it with the trend shown on the given table.

e. Predict the number of households in the United States for 2010. Interpret your answer.

4. The bottlenose dolphin is the most frequently seen along the shores of the United States. At birth, bottlenose dolphins are 42 to 52 inches long and weigh about 44 pounds. Baby dolphins (calves) nurse for more than a year and stay with their mothers for 3–6 years learning how to catch fish. Dolphins can reach seven times their birth weight in the first year. Bottlenose dolphins typically weigh 440 to 600 pounds (large dolphins can be a little over 1,000 lb) and reach an average size of 10 feet (some reach as much as 14 feet).
Sources: http://seaworld.org; http://monkeymiadolphins.org; http://www.hpu.edu/index.cfm

a. Use linear interpolation to estimate the weight of a 3-year-old large dolphin.

b. Extrapolate to predict the weight of the dolphin at 20 years of age. Is your prediction reasonable? Why or why not?

5. Professor Ludwig wants to see whether a student's test score is related to the number of review problems the student does before a test. The professor assigns a total of 45 review problems for each test. The table below denotes 12 students' test scores along with the number of review problems, out of the 45, that they completed.

No. of problems completed	42	40	34	12	45	29	36	38	43	20	25	37
Test score (%)	82	76	65	55	90	84	85	71	98	58	64	88

a. Give an appropriate window that will best fit the data.

*X*min = *X*max = *Y*min = *Y*max =

b. Construct a scatterplot using the data above. Label the axes and tick marks according to the data.

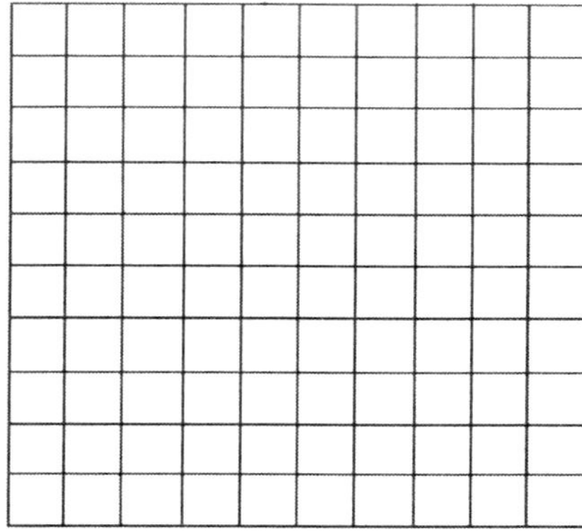

c. Find the equation of the regression line that best fits the data.

d. What score could Professor Ludwig predict if a student completes 39 review problems? Round to the nearest percent.

e. What number of review problems could he predict if the student's test score was 85%? Round to the nearest whole number.

Take it further with your grapher!

f. Determine the correlation coefficient.

g. Observe the correlation coefficient. Does this linear regression line seem like a good predictor? Explain why or why not.

6. Hedo Turkoglu has been playing for the Orlando Magic from the 2004 season through the 2008 season. A 3 pointer—that is, a 3 point field goal—is a shot that is made from outside the 23.9 foot arc from the basket. The table below displays the number of Turkoglu's 3 pointers made during his NBA (National Basketball Association) career from 2000 to 2007.

Season	No. of 3 pointers
2000–2001	28
2001–2002	63
2002–2003	29
2003–2004	101
2004–2005	93
2005–2006	114
2006–2007	109

Source: http://nba.com

a. Letting x denote the years from 2000 and y denote the number of 3 pointers, give an appropriate window using the data above.

*X*min = *X*max = *Y*min = *Y*max =

b. Create a scatterplot for the given data (label the axes and tick marks accordingly).

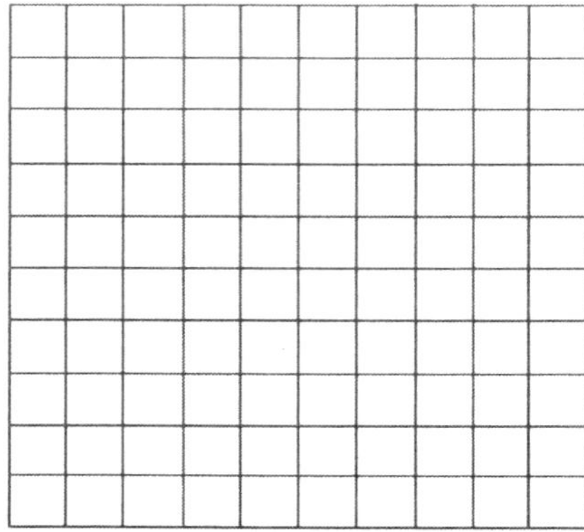

c. Find the equation of the regression line that best fits the data.

d. Graph the regression line along with the scatterplot.

e. Using the equation of the regression line, find Hedo Turkoglu's predicted successful 3 pointers for the next season.

Take it further with your grapher!

f. State the correlation coefficient.

g. Observe the correlation coefficient and your answer in part (f). Does this seem like a good way to predict his future 3 pointers? Explain why or why not.

II. Linear Systems and Applications

Linear Systems: Graphical Solutions

1. Solve the linear system using the given graph.

$$y = \frac{x+1}{2}$$
$$y = x - 1$$

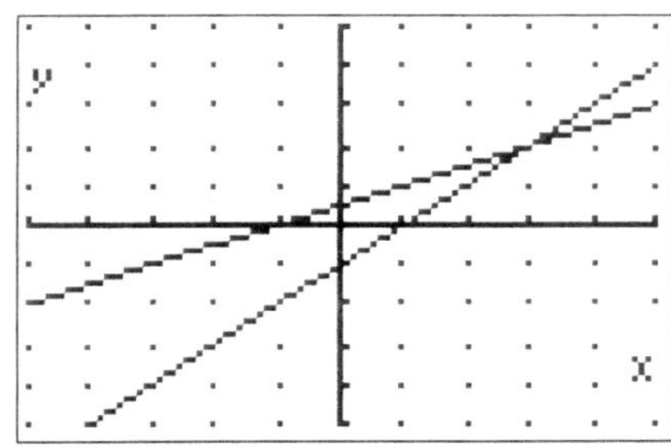

X min = -5, X max = 5, X scale = 1
Y min = -5, Y max = 5, Y scale = 1

Answer: ___________

2. The following graph shows the solution to a system of equations. Although we do not know the equations for this system, we can still find its solution with the use of the graph. Explain how.

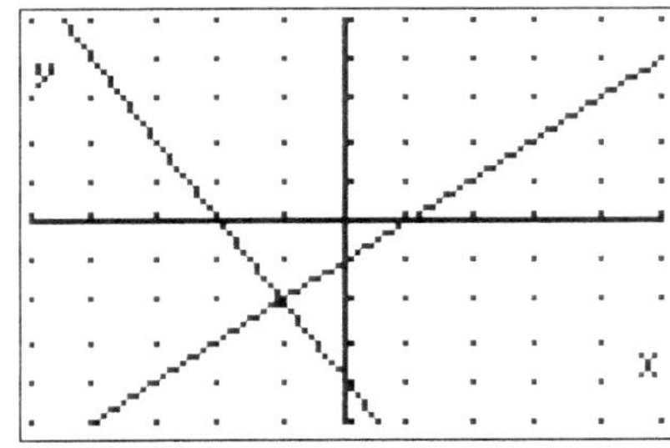

X min = -5, X max = 5, X scale = 1
Y min = -5, Y max = 5, Y scale = 1

Explain: ____
____
____

3. Solve the following linear systems *graphically*. For each problem, include a rough sketch of the graph that you created on your calculator. *Do not solve the systems algebraically*. (Hint: Remember to choose an appropriate viewing window.)

a. $y - 1 = -0.25x$
$y = 2$

Show your window here:

X min =
*X*max =
*X*scl =
*Y*min =
*Y*max =
*Y*scl =
*X*res = 1

Answer: ___________

Include a rough sketch of your graph here:

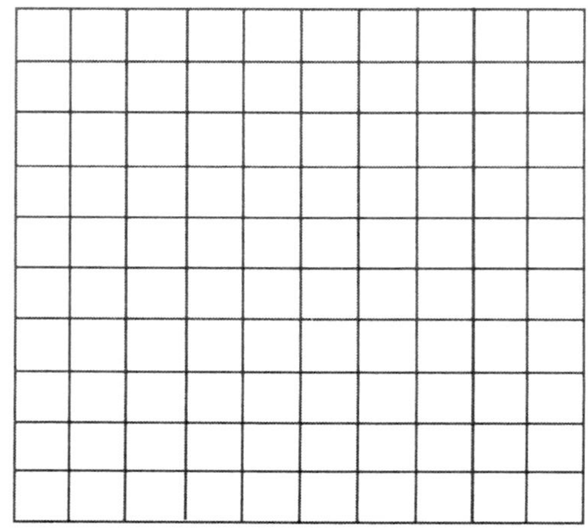

b. $50y = 33x - 7$
$7y = -3.75x + 25$

Show your window here:

*X*min =
*X*max =
*X*scl =
*Y*min =
*Y*max =
*Y*scl =
*X*res = 1

Answer: ___________

Include a rough sketch of your graph here:

4. Give a graphical example of a system of equations that is *consistent and independent*. Label the graphs as y_1 and y_2 and write their equations in slope-intercept form or in standard form. Remember to label your axes and the tick marks! You may use a different scale value other than 1. (Answers may vary from those of your classmates.)

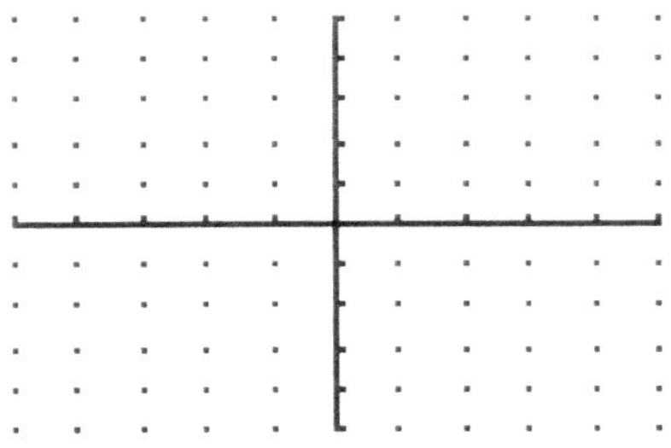

Equation for y_1 ______________________

Equation for y_2 ______________________

5. Give a graphical example of a system of equations that is *inconsistent and independent*. Label the graphs as y_1 and y_2, and write their equations in slope-intercept form or in standard form. Remember to label your axes and the tick marks! You may use a different scale value other than 1. (Answers may vary from those of your classmates.)

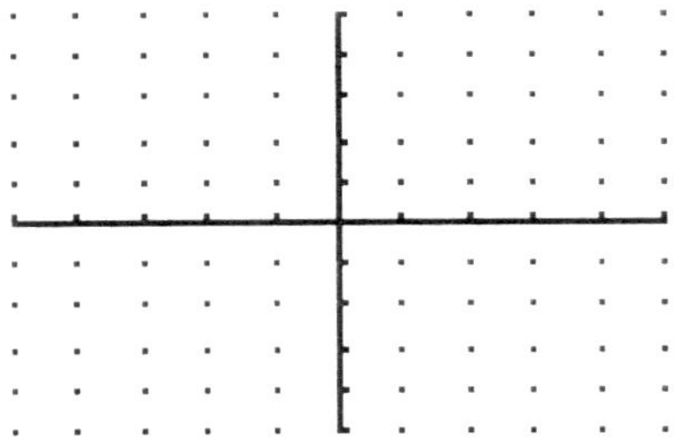

Equation for y_1 ______________________

Equation for y_2 ______________________

Linear Systems: Graphical Solutions (Applications)

1. Sidra has \$80 in her bank account and will save \$10 per week. Kurt has \$200 in his bank account and is withdrawing \$20 per week.

a. Write an equation that represents Sidra's balance in her bank account for x weeks.

b. Write an equation that represents Kurt's balance in his bank account for x weeks.

c. Solve the system of equations graphically to answer the following: When will they have the same amount of money, and how much will each have? Include the graph. (Be sure to label the axes in terms of the problem and show your chosen scale.)

Graph:

Answer:
They will have the same amount of money, \$ ______, after______ weeks.

Problem 2 deals with supply and demand, one of the most fundamental concepts of economics. Here is brief background information:

The quantity of products available for selling is called the *supply*. It is the total amount of a specific product or service that is available to consumers. The quantity of products that consumers are willing to buy is called the *demand*. It is the consumer's willingness to pay a price for a specific product or service. In other words, supply represents how much the market can offer, and demand represents how much (quantity) of a product or service buyers desire. A lower price of the products or services tends to increase demand, and a higher price tends to decrease demand. The balancing effect of supply and demand is called *equilibrium*. This occurs when a price is found that makes supply and demand equal.
Sources: http://investopedia.com; http://netmba.com

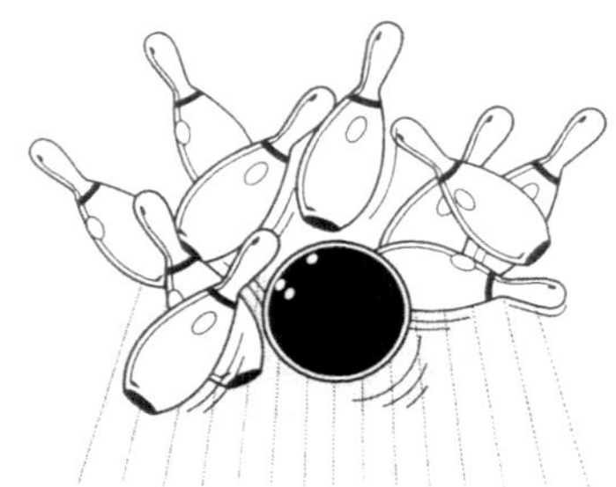

2. A new bowling magazine sells for $3.95 per issue. The printing manager studies production costs and determines that the company can continue generating 100,000 copies at the current price. The publisher of the bowling magazine wishes to print and sell more copies, and she proposes increasing the price per issue to $5.45 in order to produce 160,000 copies. The magazine marketing manager informs the publisher that the company could sell 120,000 issues at the current price, if these were available, but if the price increase is too high, for example $4.95, the public would only buy around 90,000 copies.

a. Find two data points for the supply function.

b. Find the slope between the two data points for the supply function.

c. Find a linear model for the supply function. (Hint: Use the point-slope form of a line.)

d. Find two data points for the demand function.

e. Find the slope between the two data points for the demand function.

f. Find a linear model for the demand function.

g. Use an appropriate viewing window to graph both equations. Include a rough sketch of your graph and state your window.

h. Use your graph to find the equilibrium point and state its meaning.

i. Find the equilibrium point algebraically and verify your answer from part (h).
(Remember: Equilibrium price $\Rightarrow$ Supply = Demand)

<u>Linear Systems: Numerical Solutions</u>

1. Use the following table to solve this linear system:

$y = 5x + 8$
$y = -3(x - 12)$

X	Y1	Y2
1.000	13.000	33.000
1.500	15.500	31.500
2.000	18.000	30.000
2.500	20.500	28.500
3.000	23.000	27.000
3.500	25.500	25.500
4.000	28.000	24.000
X=1		

Answer: ____________

2. The following table shows the solution to a system of equations. Although we do not know the equations for this system, we can still find its solution with the use of the table. Explain how.

X	Y1	Y2
-11.00	-34.00	-46.00
-10.00	-32.00	-40.00
-9.000	-30.00	-34.00
-8.000	-28.00	-28.00
-7.000	-26.00	-22.00
-6.000	-24.00	-16.00
-5.000	-22.00	-10.00
X= -11		

Explain: ____

____

____

3. Solve the following linear systems *numerically*. For each problem, show the table you created on your graphing calculator, and state the solution. (Hint: When solving an equation numerically, you may have to scroll up or down your table to find the solution.)

a. $y + 6 = 0.53x$
$y = 0.8x - 7.35$

Let ΔTbl = 1

X	*Y1*	*Y2*

Answer: ___________

b. $y = \frac{2}{5}(x - 6.75)$

$y + \frac{1}{10}x = 1.05$

Let ΔTbl = 0.5

X	*Y1*	*Y2*

Answer: ___________

Linear Systems: Numerical Solutions (Applications)

1. The table below gives the quantity of biology workbooks demanded and supplied for selected prices.

Price ($)	Quantity demanded	Quantity supplied
45	200	0
85	120	120
125	40	240

a. Comment about the quantity demanded and supplied as the price of the workbook increases.

b. Is the slope of the demand equation positive or negative? Find the slope and write it in fraction form.

c. Is the slope of the supply equation positive or negative? Find the slope and write it in fraction form.

d. Use the table to find the linear equation that gives the price as a function of the quantity demanded, and the equation that gives the price as a function of the quantity supplied. Be careful when identifying your x and y values—that is, the p and q values.

Demand equation: _______________ Supply equation: _______________

e. What is the quantity of books demanded and supplied at market equilibrium?

f. What is the price market equilibrium?

2. A soccer team plans to make homemade bracelets to raise money for team shirts. The players plan to sell the bracelets for $15 each. A parent helps them calculate their cost and realizes that their fixed cost is $30 for the beads, and it will cost $10 to make each bracelet.

a. Write the soccer team's revenue function, $R(x)$.

b. Write the soccer team's cost function $C(x)$.

c. Use your grapher to complete the table below.

x No. of bracelets made	$R(x)$	$C(x)$
2		
3		
4		
5		
6		
7		
8		

d. Using the table from part (c), find the break-even point.

e. What is the price at the break-even point?

f. Graph the revenue and the cost functions below. Label the equations as $R(x)$ and $C(x)$ and identify the break-even point. Use the given window.

Xmin = 0

Xmax = 10

Xscl = 1

Ymin = 0

Ymax = 150

Yscl = 10

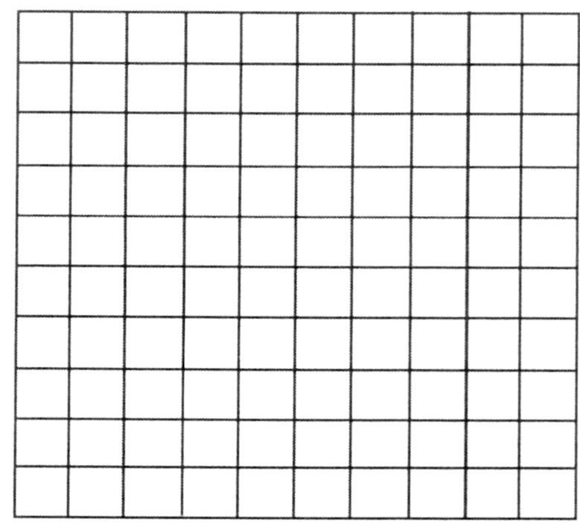

Linear Systems: Substitution and Elimination

1. Luis and Phil decide to go to a hot air balloon race. On their way to the race, they decide to bet on which balloon they think will win. Luis bets on Team Balloonza, and Phil bets on the Balloon Grand Nationals. Once they get to the race, they realize that it has already started. Speaking with another spectator, they find out that Team Balloonza is 95 feet above the ground and rising at 40 feet per minute. The balloon of the Grand Nationals is only 30 feet above the ground but is rising at 50 feet per minute.

a. Write an equation for the height of each balloon as a function of time.

Team Balloonza ________________ Balloon Grand Nationals ______________

b. Use substitution to find when the balloons will be at the same height.

c. What will their height be when they are tied in the race?

Team Balloonza

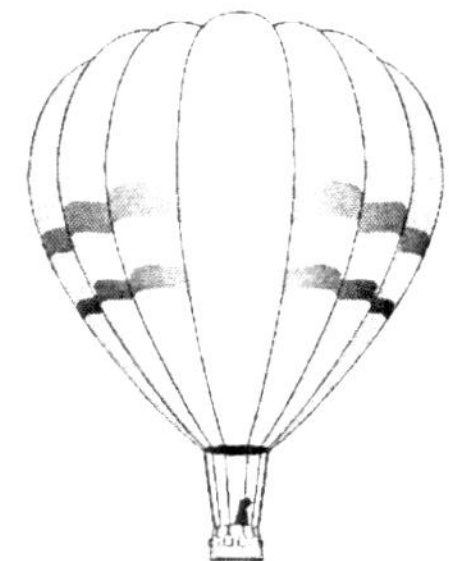

Balloon Grand Nationals

2. Chaoxiang had a total of $2,964 in his checking account and money market account. In 2008, his money market account paid 2.64% in interest, and his checking account paid 1.88% in interest. If he earned $70.31 in interest for that year, how was his money divided between the two accounts?

a. Write an equation about the total amount of money in both accounts.

b. Write an equation about Chaoxiang's annual interest from both accounts.

c. Solve the system algebraically and answer the question in the problem. Write your answer in a complete sentence. Set up your calculator to 4 decimal places.

3. Your community college's theater company is presenting a hip hop musical during the spring term. Ticket prices for all performances are \$8 for general admission and \$6 for students, college staff, or alumni. If the theater company sells 500 tickets and collects \$3,380, how many general admission tickets and how many students, staff, and alumni tickets are sold?

a. Write an equation about the total number of tickets the company sells.

b. Write an equation for the revenue from both tickets.

c. Solve the system algebraically and answer the question in the problem. Write your answer in a complete sentence.

4. At Ana and Biagio's gourmet chocolate cookies store, 2 boxes of triple-chocolate pecan cookies and 4 boxes of super-chocolate cherry cookies cost \$177. Six boxes of triple-chocolate pecan cookies and three boxes of super-chocolate cherry cookies cost \$261. What is the price per box of each type of gourmet chocolate cookie?

a. Write a system of equations to address this problem. Use p for pecan and c for cherry.

b. Solve the system algebraically and answer the question in the problem. Write your answer in a complete sentence.

c. Did you solve this problem by using the substitution method, or did you decide to use the elimination method? Explain your decision. How much would it cost to buy 8 boxes of the pecan cookies and 12 boxes of the cherry cookies?

III. Inequalities in Two Variables

Graphical Solutions

1. The graph below has the following viewing window: [-8, 8, 2] by [-9, 9, 3]. Label the tick marks according to the given window. Label the graph with the positive slope as $f(x)$, and the graph with the negative slope as $g(x)$. Then, use the graphs to find where:

a. $f(x) = g(x)$ b. $g(x) < f(x)$ c. $f(x) \leq g(x)$ d. $f(x) > g(x)$

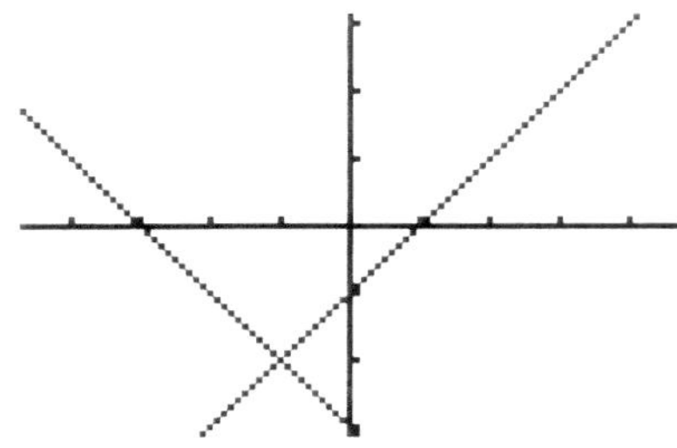

Answers:

a. __________

b. __________

c. __________

d. __________

2. Graph the following inequalities, find the solution, and identify the vertices (or corner points) of the boundary. State the role of the inequalities $x \geq 0$ and $y \geq 0$.

$3x + 4y \leq 12$
$y \geq -2x + 4$
$x \geq 0, y \geq 0$

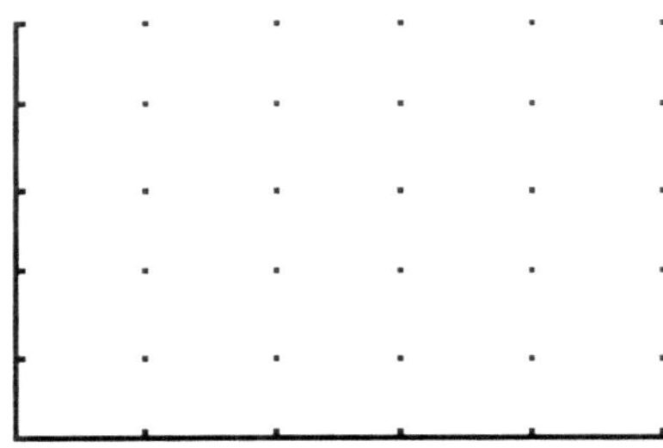

Applications

1. It takes Remy, a teaching aide, 45 minutes to prepare a math quiz and 2 hours to prepare a math project. He has been asked to develop various quizzes and projects for a faculty member. Based on his work schedule, he can spend no more than 25 hours performing both tasks together.

a. Write an inequality that represents Remy's situation, where x is the number of quizzes he prepares and y is the number of projects he develops.
   (Hint: Remember to use equivalent units of time!)

b. Graph your inequality. Use a solid line or a dotted line as appropriate. Shade the corresponding region. (Label the axes in terms of this problem and show the scale.)

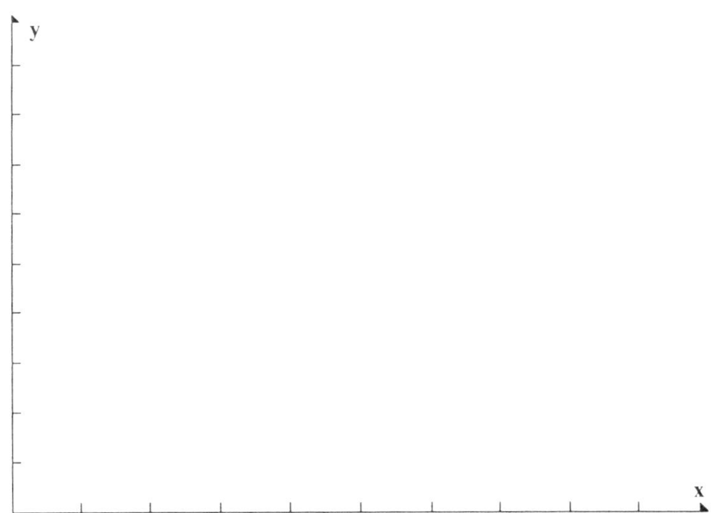

c. If Remy prepares five projects, how many quizzes can he still develop during the time he has left?

d. State two additional possible combinations (quizzes and projects) that Remy could develop.

2. After working for a couple hours, a large team of faculty members developing a new course is ready to take a coffee break. They go to a restaurant that charges \$2 for a 12-ounce café latte and \$3 for a12-ounce mocha.

a. Let x be the number of 12-ounce café lattes and y be the number of 12-ounce mochas. If the faculty members decide that they will spend less than \$30 altogether, write a linear inequality that represents how much of each type of coffee the group can buy.

b. Graph your inequality. Use a solid line or a dotted line as appropriate. Shade the corresponding region. (Label the axes in terms of this problem and show the scale.)

c. State three possible coffee combinations (café latte and mocha) the team can buy.

3. Jimmy's newspaper business has a monthly revenue of $R(x) = 145x$ and monthly cost of $C(x) = 95x + 1{,}850$, where x is the number of units produced and sold. How many newspapers must he sell in a month to make a profit? State your inequality and solve it to answer the question. Recall that Profit equals Revenue minus Cost; that is, $P(x) = R(x) - C(x)$.

Inequality ____________________________

Solution ____________________________

4. Mr. Han applies for two sales positions and is curious about how much he must sell for one position to be more desirable than the other. He does not want to take the commission position if he feels that the number of sales would be unreasonable. The management position pays a flat rate of \$4,200 per month. The sales position pays a flat rate of \$3,000 but is has a commission rate of 3% on all sales. What amount would Mr. Han need to sell in a month for the sales position to pay more than the management position?

Inequality ____________________________

Solution ____________________________

5. Naomi's handmade baskets enterprise has monthly profits from basket sales as $P(x) = 7.85x - 942$ dollars, where x is the number of baskets sold.

a. How many baskets must she sell to break even?

b. How many must be sold to avoid a loss?

6. According to medical experts, an adult's LDL (bad) cholesterol level should be less than 100. An adult's HDL (good) cholesterol level should be between 40 and 60.
Source: http://americanheart.org

a. Write an inequality for the recommended LDL level.

b. Write an inequality for the recommended HDL level.

c. Construct a graph that shows the appropriate cholesterol levels for an adult. Let x represent the LDL and y represent the HDL.

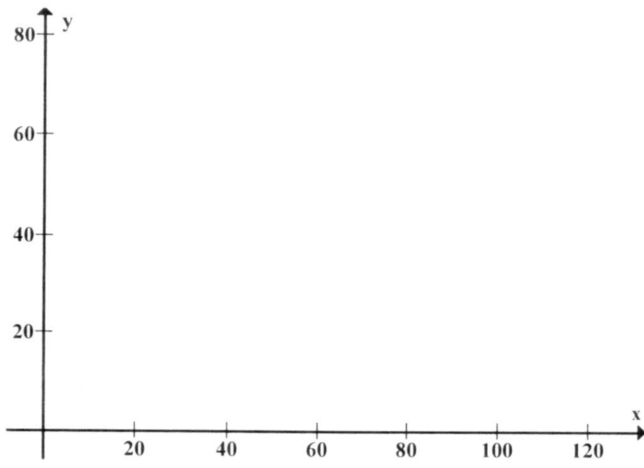

IV. Non-Linear Functions and Applications

Basic Functions and Their Graphs

1. Each of the following is a variation of the graph of a basic function. Identify the basic function in each case.

a. 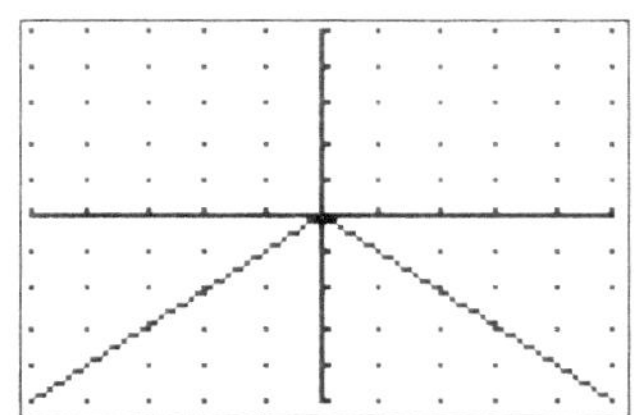

$f(x) =$ _______________

b.

$f(x) =$ _______________

c. 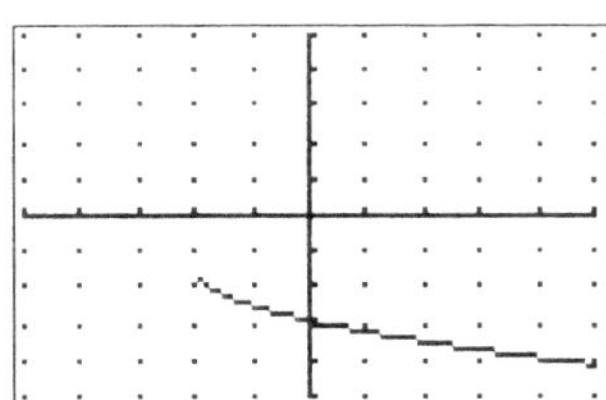

$f(x) =$ _______________

d. 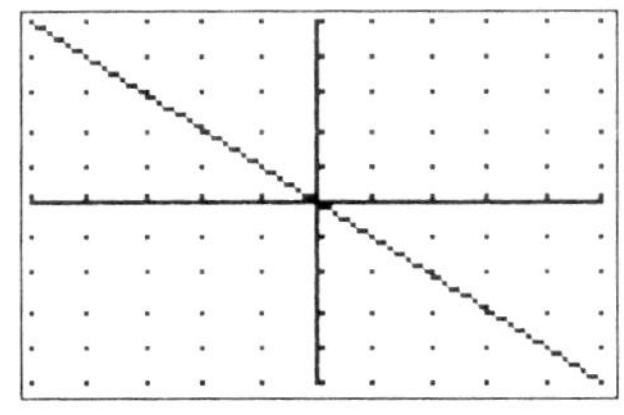

$f(x) =$ _______________

e. 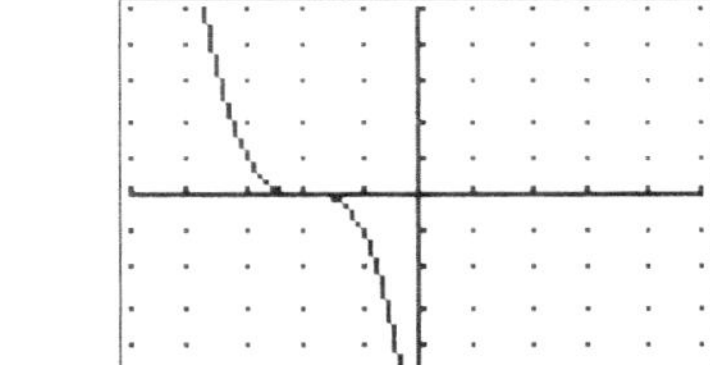

$f(x) =$ _______________

f.

$f(x) =$ _______________

g.

$f(x) =$ _______________

h. 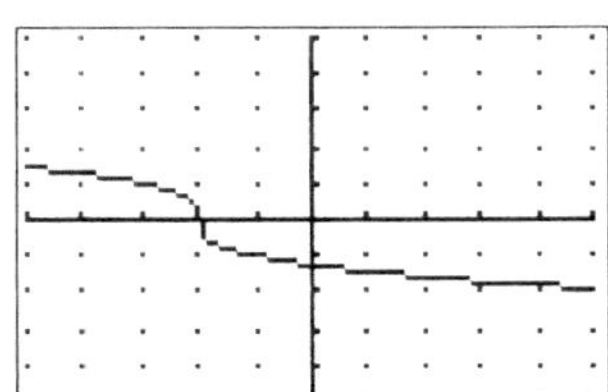

$f(x) =$ _______________

2. The figure below illustrates the graph of $f(x) = 2x^3 + 4x^2 - 2x - 3$. Use the graph to answer the questions.

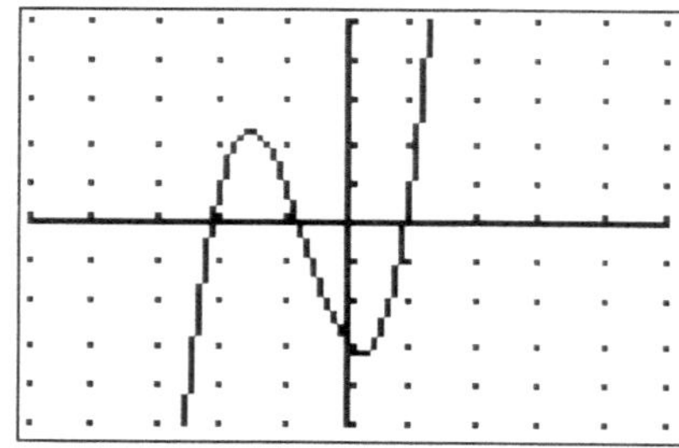

X min = -5, X max = 5, X scale = 1
Y min = -5, Y max = 5, Y scale = 1

a. Identify the basic function associated with this function.
b. Find $f(0)$.
c. What do the coordinates found on (b) represent?
d. Find the zeros of the function.
e. Find x so that $f(x) = -4$.
f. Solve: $2x^3 + 4x^2 - 2x - 3 = 1$
g. Solve: $2x^3 + 4x^2 - 2x - 3 > 0$
h. Solve: $2x^3 + 4x^2 - 2x - 3 \leq 0$

Answers:

a. ______________________________

b. ______________________________

c. ______________________________

d. ______________________________

e. ______________________________

f. ______________________________

g. ______________________________

h. ______________________________

3. Graph the following sets of functions with your graphing calculator and explain their similarities and differences. You may use the standard window. Include a rough sketch for each graph.

a. $f(x) = \|x\|$	b. $f(x) = \sqrt{x}$	c. $f(x) = x^2$
$f(x) = \|x + 3\|$	$f(x) = \sqrt{x+3}$	$f(x) = (x + 3)^2$
$f(x) = \|x - 3\|$	$f(x) = \sqrt{x-3}$	$f(x) = (x - 3)^2$
$f(x) = \|x\| + 3$	$f(x) = \sqrt{x} + 3$	$f(x) = x^2 + 3$
$f(x) = \|x\| - 3$	$f(x) = \sqrt{x} - 3$	$f(x) = x^2 - 3$

Answers:

a. Graphs:

Similarities:

Differences:

b. Graphs:

Similarities:

Differences:

c. Graphs:

Similarities:

Differences:

Transformations

1. If $f(x) = \frac{1}{x}$, describe each of the following transformations:

a. $f(x) = \frac{1}{x} + 50$

b. $f(x) = \frac{1}{x} - 50$

c. $f(x) = \frac{1}{x+50}$

d. $f(x) = \frac{1}{x-50}$

e. $f(x) = \frac{1}{x+50} + 50$

f. $f(x) = \frac{1}{x+50} - 50$

g. $f(x) = \frac{1}{x-50} + 50$

h. $f(x) = \frac{1}{x-50} - 50$

Answers:

a. ____

b. ____

c. ____

d. ____

e. ____

f. ____

g. ____

h. ____

2. How does a scale factor (vertical scaling) affect the graph of a function? Give an example.

3. How is the graph of $y = -f(x)$ different from the graph of $y = f(x)$? Give an example.

4. Each of the following is a transformation of the graph of a basic function. Identify the basic function in each case, and write the formula for the transformed graph.

Graph X min = -5, X max = 5, X scale = 1 Y min = -5, Y max = 5, Y scale = 1	Basic function	Formula for transformed graph
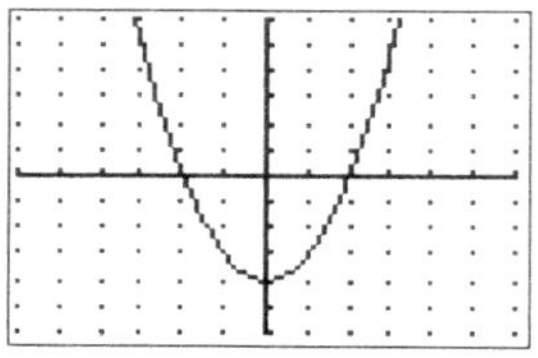		
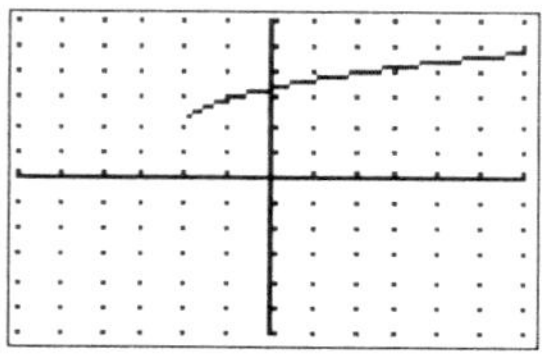		
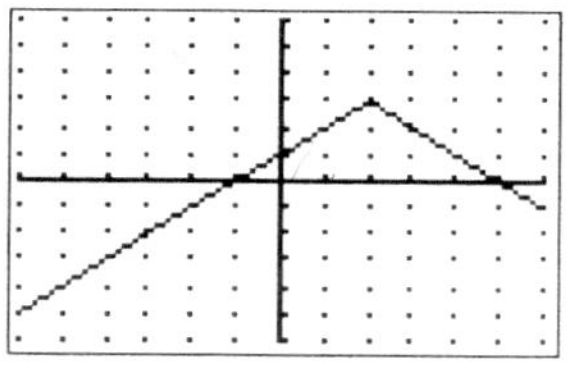		
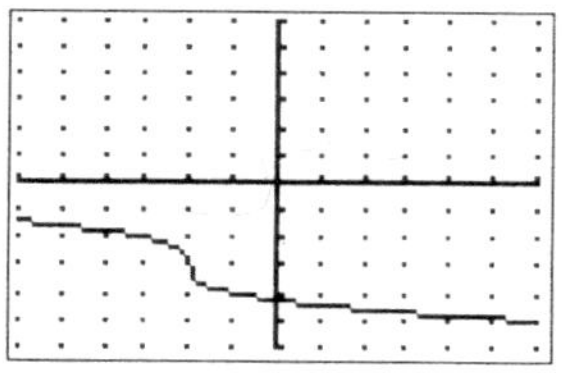		

5. Deborah was asked to compare the graphs of $y = 25|x|$ and $y = -0.25|x|$. She said the graph of $y = 25|x|$ was wider and concave up, while the graph of $y = -0.25|x|$ was narrower and concave down. Was Deborah totally right? Partially right? Or totally wrong? Explain how you decided.

6. An object is dropped from a height of 100 feet. Its height, h, (in feet) after t seconds is modeled by the following function:

$$h(t) = -16t^2 + 100 \quad \text{for } t \geq 0$$

a. Using your grapher, construct the graph of the function $h(t) = -16t^2 + 100$.

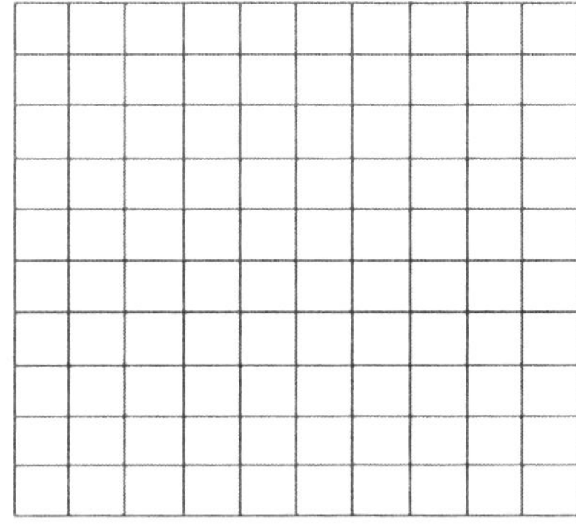

b. Describe the transformations that must occur to generate $h(t)$ from the graph of $f(t) = t^2$.

c. A second object is dropped from the same altitude, and its height is modeled by $s(t) = -16(t - 2.5)^2 + 100$. Using your grapher, construct the graph of $s(t)$.

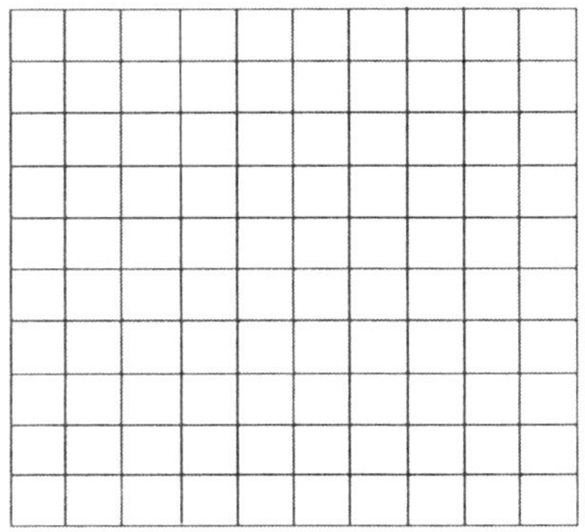

d. Interpret the transformation in part (c) in the context of this problem.

Domain and Range

1. Find the domain and range of each function algebraically; write your answer using interval notation. Graph the function with your graphing calculator and corroborate your algebraic answers; include a rough sketch of your graph.

Function	Domain	Range	Graph
$f(x) = \sqrt{x-3}$			
$f(u) = 6 - \lvert u \rvert$			
$f(u) = \lvert 6 - u \rvert$			

2. Determine the domain and range of the functions whose graphs are provided; write each answer using interval notation.

a.

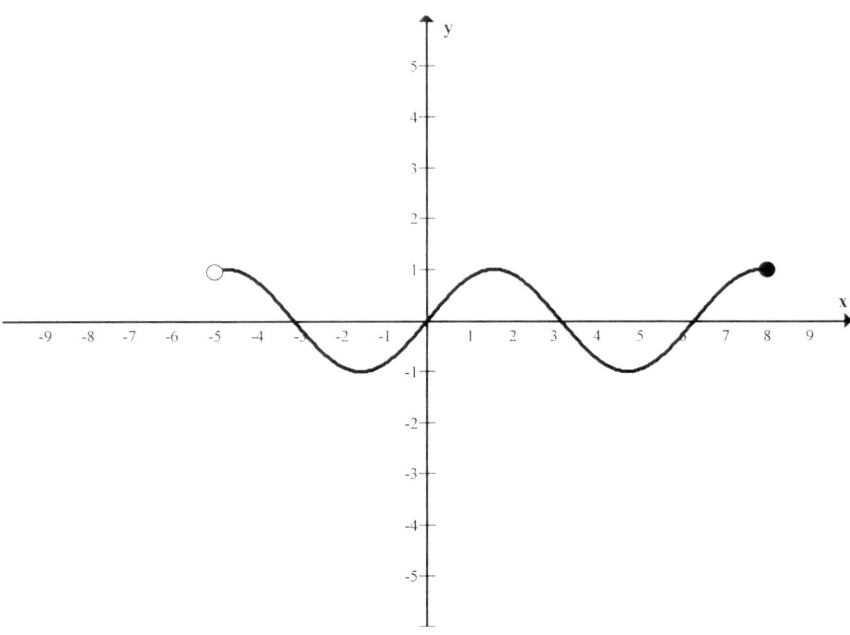

Domain ____________________ Range ____________________

b.

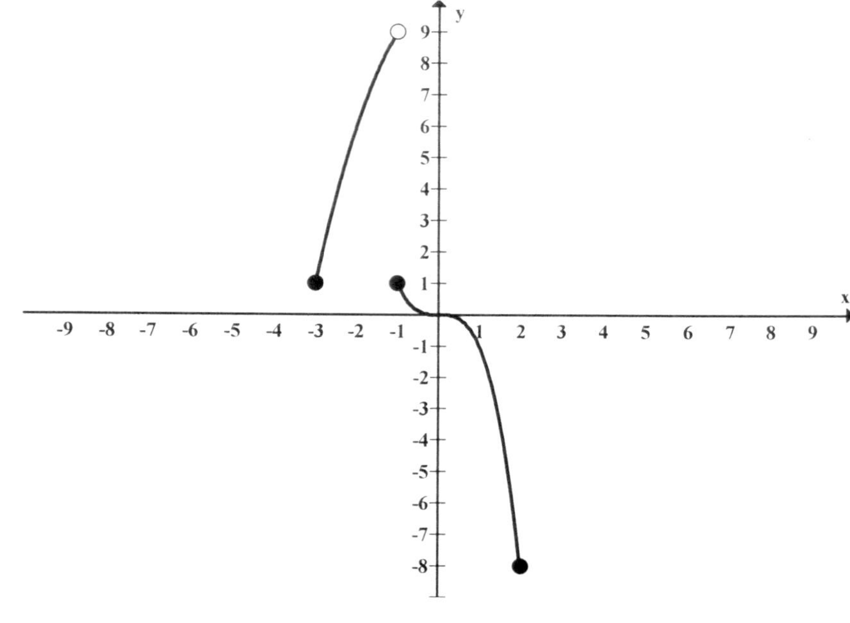

Domain ____________________ Range ____________________

3. A rocket is launched upward from the ground at 72 feet per second. Its height (in feet) is denoted by $h(t)$, where t represents time (in seconds).

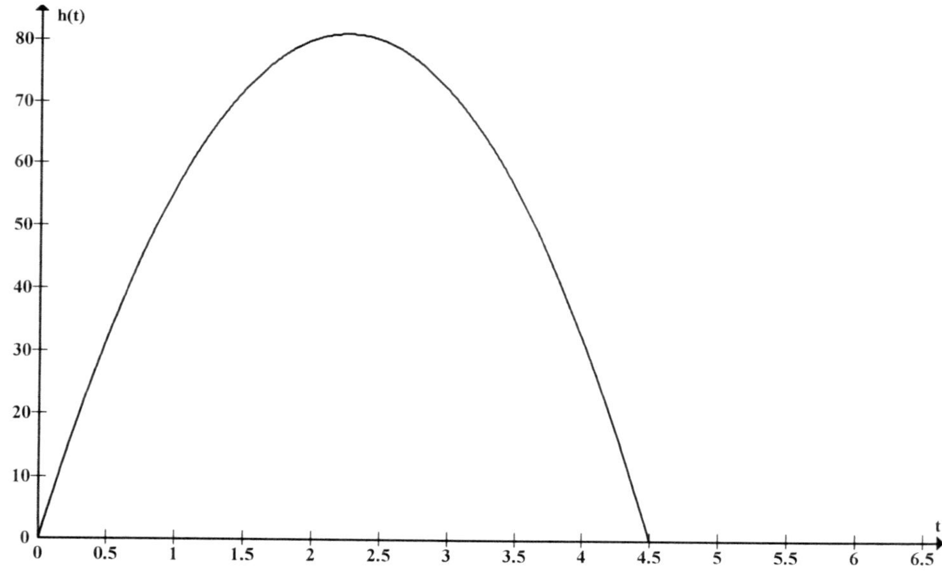

a. When does the rocket hit the ground? Provide an approximate time.

b. From the graph above, estimate the domain and the range.

4. Find the domain and range for the following graphs; if the graph does not represent a function, declare it. Assume each tick mark is 1 unit. If endpoints do not appear on the graph, assume there are arrows.

a. Domain ________________

Range ________________

b. Domain ________________

Range ________________

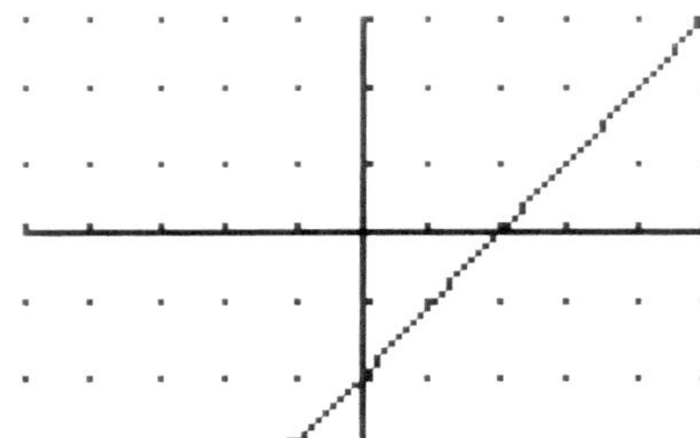

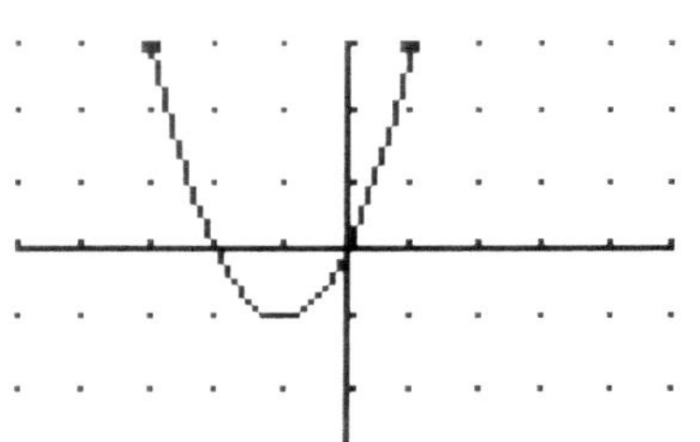

c. Domain ________________

Range ________________

d. Domain ________________

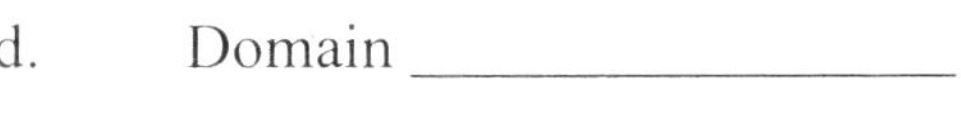

Range ________________

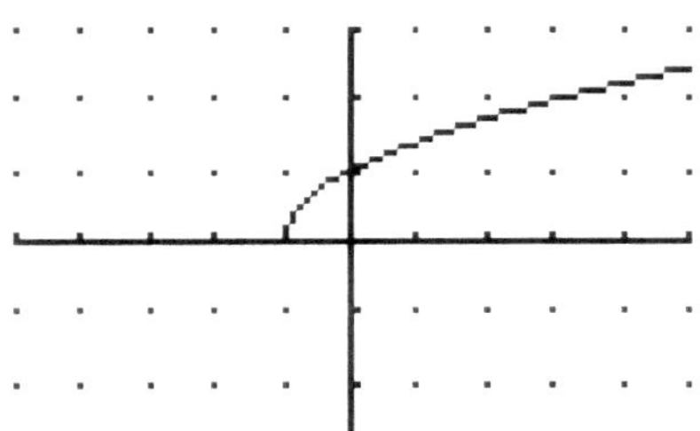

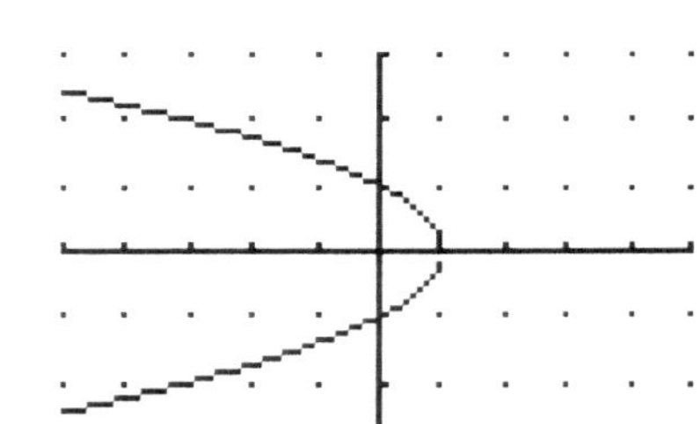

e. Domain ________________

Range ________________

f. Domain ________________

Range ________________

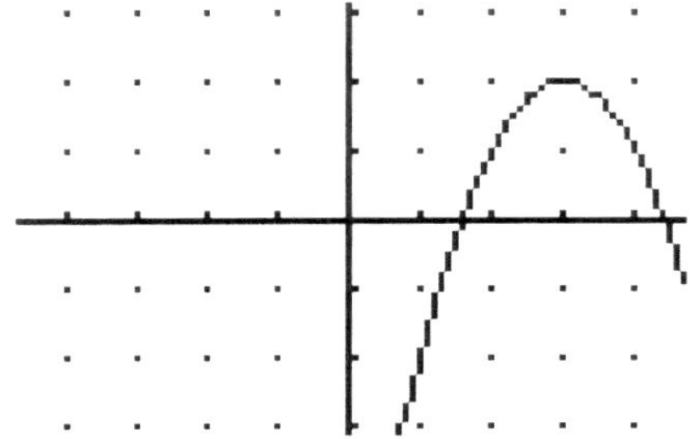

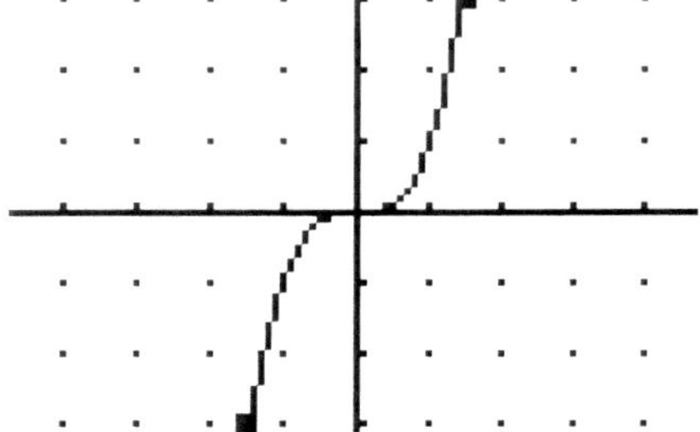

g. Domain ______________

Range ________________

h. Domain ________________

Range ________________

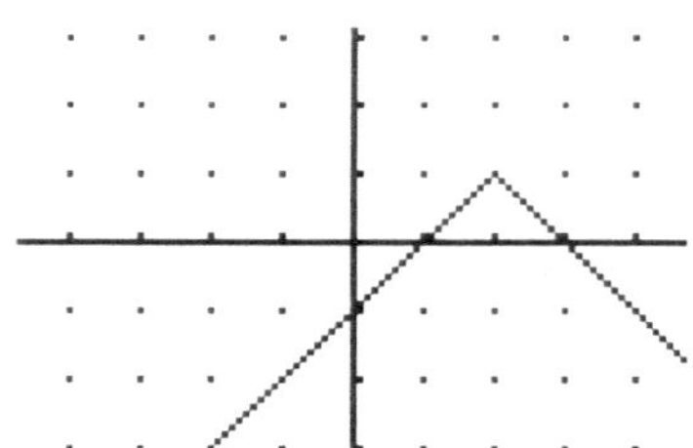

5. Each of the following functions has a restricted domain. State the restriction and explain why it occurs.

a. $f(x) = \dfrac{1}{x+50}$ b. $f(x) = \dfrac{1}{x^2 - 7x + 10}$ c. $f(x) = \sqrt{x - 35}$

Answers:

a. ____

b. ____

c. ____

6. The graphs of $f(x) = -x^2 + 3$ and $f(x) = -(x - 5)^2 + 3$ are shown on the figure below.

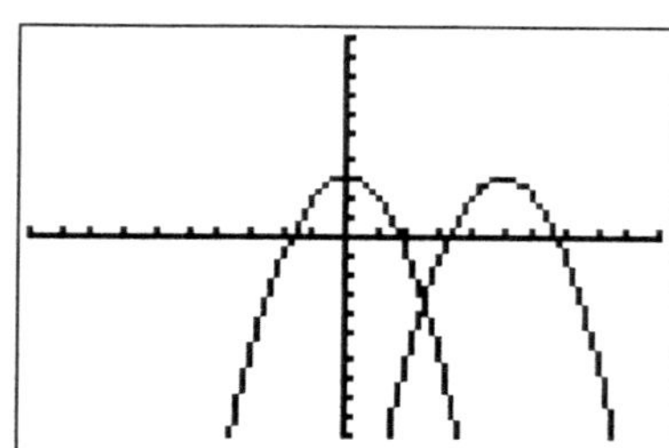

Viewing window: [-10, 10, 1] by [-10, 10, 1]

Use the figure to decide which of the following statements is *true* and explain your decision.

a. Both functions have different domains.
b. Both functions have the same range.

Answer: ____

7. A pendulum is an object suspended from a fixed point so it swings freely back and forth under the influence of gravity. It is commonly used to regulate devices such as clocks. The period of a simple pendulum is the time it takes for two swings (left to right and back again) of the pendulum—that is, the time required to complete a full cycle. The formula for the period, T, is given by the formula.

$$T = 2\pi\sqrt{\frac{l}{g}}$$

where l is the length of the pendulum in meters, and g is the acceleration due to Earth's gravity, or 9.80 m/s^2.

What restrictions, if any, would be needed on the domain of this formula so that the values obtained with it would make sense? Justify your answer.

Answer: ____
____

8. The following table shows the top-selling cookie sales ending in the year 2004. The sales are for supermarkets, drugstores, and mass merchandisers, excluding Wal-Mart.

Cookie	Nabisco Oreo	Nabisco Chips Ahoy	Keebler Chips Deluxe	Nabisco Newtons	Pepperidge Farm Distinctive
Sales	449,772,768	317,724,064	115,821,328	119,715,920	131,371,384

Source: World Almanac 2005

a. Identify the domain.

b. Identify the range.

c. Which company seems to sell the most cookies, and what were its total sales for the year 2004?

9. The following table shows the population of Florida for the indicated years.

Year	1980	1990	2000	2003	2007
Florida Population	9,746,961	12,937,926	15,982,378	17,019,068	18,251,243

Sources: http://www.census.gov; http://www.stateofflorida.com

a. Explain why this table represents a function.

b. The two variables are year and Florida population. Which variable corresponds to the independent variable and which to the dependent variable, and *why*?

c. Identify the domain and range of the function.

d. Comment on Florida's population growth trend since 1980.

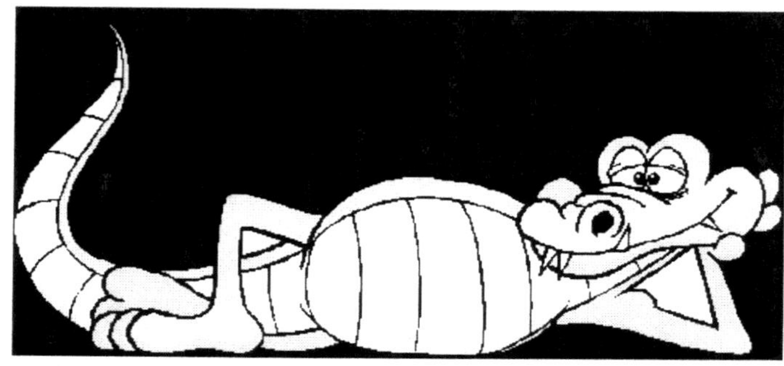

10.

Function	Graph Label the axes!	State the function domain and range	Provide an example of a vertical shift	Provide an example of a horizontal shift	Provide an example of a reflection	Provide an example of a vertical stretch	Provide an example of a vertical compression
$f(x) = x^2$		Domain: $(-\infty, \infty)$ Range: $[0, \infty)$	Function: $f(x) = x^2 + 3$ Graph:	Function: $f(x) = (x + 3)^2$ Graph:	Function: $f(x) = -x^2$ Graph:	Function: $f(x) = 8x^2$ Graph:	Function: $f(x) = 0.08x^2$ Graph:
$f(x) = \sqrt{x}$		Domain: Range:	Function: Graph:	Function: Graph:	Function: Graph:	Function: Graph:	Function: Graph:
$f(x) = x^3$		Domain: Range:	Function: Graph:	Function: Graph:	Function: Graph:	Function: Graph:	Function: Graph:
$f(x) = \sqrt[3]{x}$		Domain: Range:	Function: Graph:	Function: Graph:	Function: Graph:	Function: Graph:	Function: Graph:

10.

Function	Graph Label the axes!	State the function domain and range	Provide an example of a vertical shift	Provide an example of a horizontal shift	Provide an example of a reflection	Provide an example of a vertical stretch	Provide an example of a vertical compression
$f(x) = \frac{1}{x}$		Domain: Range:	Function: Graph:	Function: Graph:	Function: Graph:	Function: Graph:	Function: Graph:
$f(x) = \frac{1}{x^2}$		Domain: Range:	Function: Graph:	Function: Graph:	Function: Graph:	Function: Graph:	Function: Graph:
$f(x) = \lvert x \rvert$		Domain: Range:	Function: Graph:	Function: Graph:	Function: Graph:	Function: Graph:	Function: Graph:
$f(x) = x$		Domain: Range:					

Source: Functions Matrix (adaptation). Permission by Mr. Lee Graubner

Piecewise Functions

1. Anthony has a full-time job as a math tutor at a local tutoring center. He gets paid \$12 an hour for 40 hours per week. During final exams week, his supervisor allows him to work overtime. When Anthony works more than 40 hours, he gets paid 1.5 times his normal hourly wage.

a. Write a function that denotes his pay, assuming he does not work more than 40 hours.

b. The computation below describes Anthony's pay when he works overtime. Let $h - 40$ denote the hours he works overtime. For example, if he works 45 hours then $45 - 40 = 5$ hours of overtime.

$12(40) + 12(1.5)(h - 40) = 480 + 18(h - 40) = 480 + 18h - 720 = 18h - 240$

Therefore, the piecewise function for his pay is described by:

$$p(h) = \begin{cases} 12h & \text{if } 0 \le h \le 40 \\ 18h - 240 & \text{if } h > 40 \end{cases}$$

Graph the piecewise function that gives his weekly pay, p, in terms of hours, h, Anthony works.

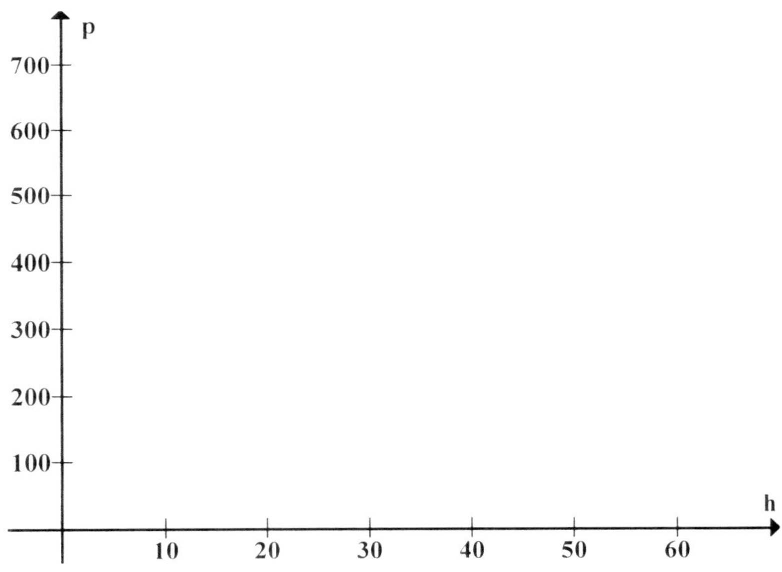

c. How much will Anthony earn if he works 46 hours?

2. Al wants to buy his soccer team shirts with their team name and the player's last name on the back. He is considering buying the shirts for his team and asking the other teams if they would also buy from the same company to reduce the per-unit cost of the shirts. Scott's Silk Screen Shop charges an initial charge of $40 to create the team logo and names. The cost per shirt is $20 for orders of 30 or less. If more than 30 shirts are ordered, the cost is only $15.

a. Write a piecewise function that gives the cost, C, of the order for x number of shirts.

b. What is the total cost for the shirts if Al orders 20 shirts for his team?

c. What would Al's total cost be if his team and another team order a total of 42 shirts?

d. How much money would Al save with the other team ordering from the same company?

e. Graph your piecewise function.

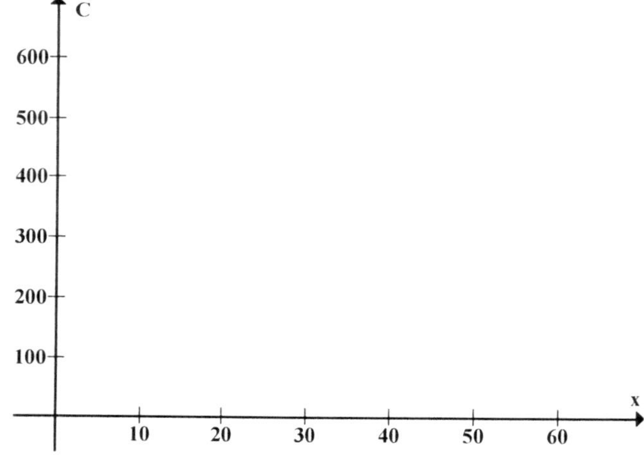

3. Sprint Nextel offers different individual wireless calling plans. One of them costs \$39.99 per month for 450 anytime minutes, excluding regulatory charges (e.g., taxes, surcharges, and other fees). The plan charges \$0.45 per additional anytime minute.
Source: http://sprint.com

The function below represents the monthly cost, C, (excluding regulatory charges) for x number of anytime minutes used.

$$C(x) = \begin{cases} 39.99 & \text{if } 0 \le x \le 450 \\ 0.45x - 162.51 & \text{if } x > 450 \end{cases}$$

a. Graph this piecewise function *by hand*. Consider the restrictions on x for each piece (sub-domain).

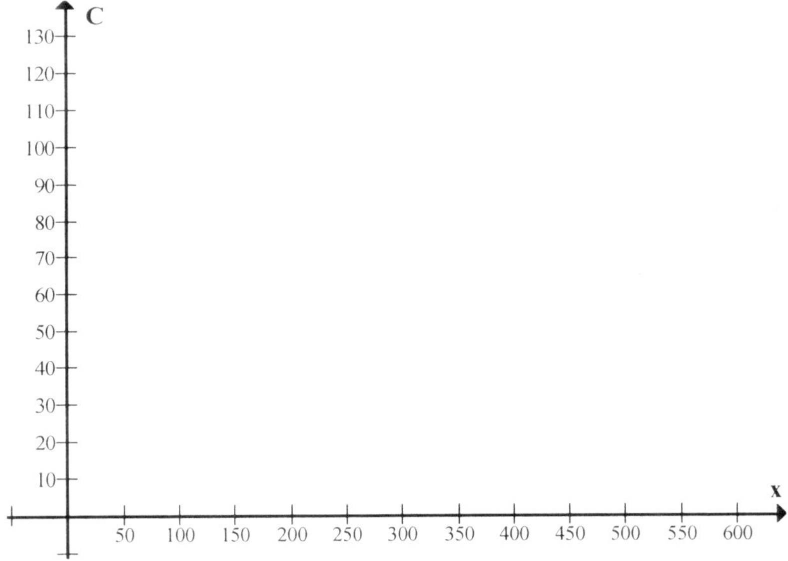

b. Calculate and interpret $C(300)$.

c. Calculate and interpret $C(550)$.

4. The federal tax rate schedule for single individuals filing their 2007 income tax return is given by the following table:

If the taxable income is over...	But not over...	The base tax is...	Plus...	Of the amount over...
\$0	\$7,825	\$0	10%	\$0
\$7,825	\$31,850	\$782.50	15%	\$7,825
\$31,850	\$77,100	\$4,386.25	25%	\$31,850
\$77,100	\$160,850	\$15,698.75	28%	\$77,100
\$160, 850	\$349,700	\$39,148.75	33%	\$160,850
\$349,700	No limit	\$101,469.25	35%	\$349,700

Source: http://www.irs.gov

For example: A single individual with a taxable income of \$50,000 would fall under the 25% bracket. The estimated taxes for 2007 for this person are calculated as follows:

The base tax, \$4,386.25, plus 25% of the amount over \$31,850.
The amount over \$31,850 is: \$50,000 – \$31,850 = \$18,150
25% of \$18,150 = \$4,537.50
The estimated tax for this individual is: \$4,386.25 + \$4,537.50 = \$8,923.75

a. Let x equal the taxable income (in dollars) for single individuals filing their 2007 income tax return. Let $f(x)$ be the tax owed to the federal government (in dollars).
Complete the piecewise function, $f(x)$, that models the federal income tax rate for 2007:

$$f(x) = \begin{cases} .10x & \text{if } 0 < x \leq 7{,}825 \\ 782.50 + .15(x - 7{,}825) & \text{if } 7{,}825 < x \leq 31{,}850 \\ 4{,}386.25 + .25(x - 31{,}850) & \text{if } 31{,}850 < x \leq 77{,}100 \\ 15{,}698.75 + .28(x - 77{,}100) & \text{if } 77{,}100 < x \leq 160{,}850 \\ \rule{6cm}{0.4pt} \\ \rule{6cm}{0.4pt} \end{cases}$$

b. Use the function to find $f(40{,}000)$.

c. Interpret your answer from part (b).

d. Solve for x: $782.50 + .15(x - 7{,}825) = 2{,}908.75$

e. Interpret your answer from part (d).

5. Discovery problem.

a. Graph the following piecewise function *by hand*. Remember to consider the restrictions on x for each piece (sub-domain).

$$f(x) = \begin{cases} |x| & if \;\; x \geq 2 \\ -x^2 & if \;\; x < 2 \end{cases}$$

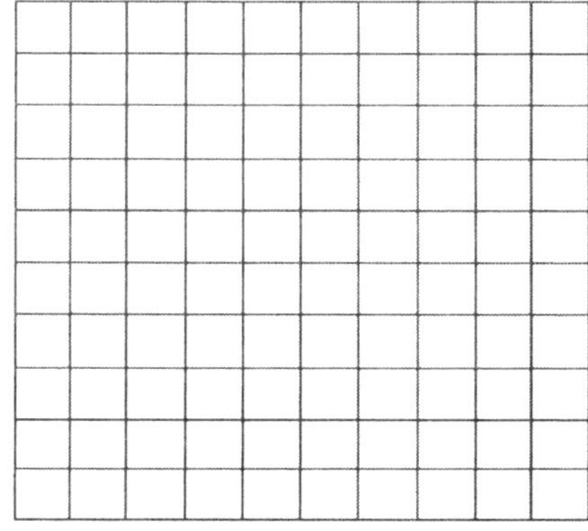

b. If we were to graph the same function with the graphing calculator, we would enter the first piece under Y_1 and the second piece under Y_2, and the graph would show as follows:

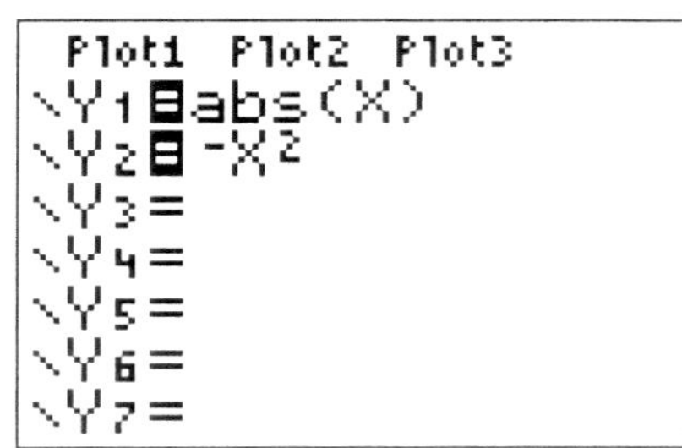

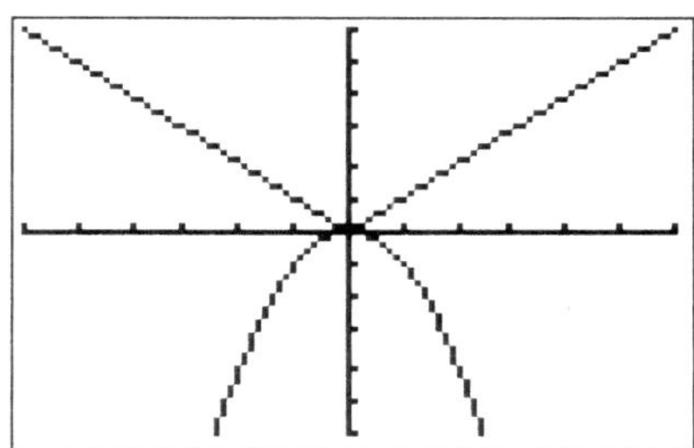

This graph no longer reflects a function. Explain why.

____

____

c. Enter the first piece of the function, $f(x) = |x|$, under Y_1 on your graphing calculator, adhering to the following keystrokes:

$Y_1 =$ [MATH] [▶] [1] [X,T,Θ,*n*] [)] ÷ [(] [X,T,Θ,*n*] [2nd] [MATH] [4] [2] [)] [GRAPH]

Explain what happened to the graph of $f(x) = |x|$.

____

____

d. Enter the second piece of the function, $f(x) = -x^2$, under Y_2 on your graphing calculator, adhering to the following keystrokes:

$Y_2 =$ - [X,T,Θ,n] [x^2] ÷ [(] [X,T,Θ,n] [2nd] [MATH] [5] [2] [)] [GRAPH]

Explain what happened to the graph of $f(x) = -x^2$.

____

____

e. If you followed the keystrokes on parts (c) and (d), your calculator will correctly display the graph of the piecewise function from part (a). It should look similar to the following:

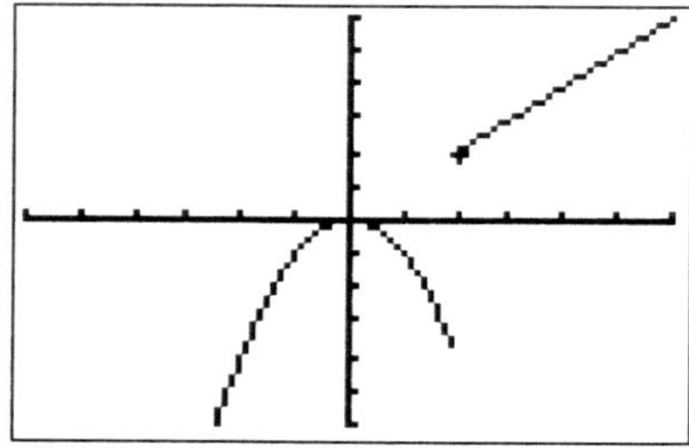

Based upon your discovery from parts (c) through (e) and using the appropriate keystrokes, graph the following piecewise function with your grapher, considering the restrictions on x for each piece (sub-domain). Make sure your graph reflects a function (apply the vertical line test). Show your keystrokes and make a rough sketch of your graph.
(Hint: Remember to use sets of parentheses when needed.)

$$f(x) = \begin{cases} x & if \ \ x \le -2 \\ 3 & if \ \ x > -2 \end{cases}$$

Write your keystrokes here:

Show a sketch of the graph displayed by your graphing calculator:

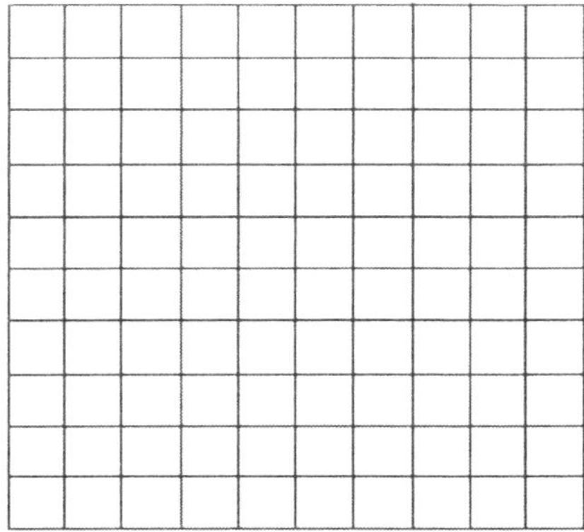

f. Summarize how you can appropriately graph a piecewise function with the graphing calculator.

____

____

Note: When graphing with the calculator, a 3-part inequality must be expressed separately. Example: For $0 \le x \le 5$, you would need to enter this as two separate inequalities: $x \ge 0$ and $x \le 5$; the "and" conjunction can be accessed by typing
[2nd] [MATH] [▶]

Variation

1. Classify each of the following as a direct or inverse variation real-life situation.

a. A construction worker who is paid hourly knows that working longer means making more money.

Circle one: Direct variation Inverse variation

Explain: ______________________________

b. A professional bowler must travel to many cities across the nation to compete in high-money tournaments. The bowler knows that driving 80 mph (if allowed by law) will get him/her to the next tournament faster than driving 70 mph—that is, the athlete will spend less time on the road.

Circle one: Direct variation Inverse variation

Explain: ______________________________

c. In car racing, a pit stop is when the car stops during a race for refueling, changing tires, repairs, or other reasons. Selecting wisely the number of pit stops is crucial in having a successful race. An unscheduled or long pit stop can be costly for a driver's chance of winning, because while his/her car is halted, the other cars on the track can rapidly gain advantage over the stopped car. The racecar driver knows that finishing 70 laps before making a pit stop is better than completing only 50.

Circle one: Direct variation Inverse variation

Explain: ______________________________

d. At a birthday celebration, each guest will receive a larger slice of cake if there are fewer attendees and less if there is a greater number of guests.

Circle one: Direct variation Inverse variation

Explain: ______________________________

2. The following graph represents a direct variation.

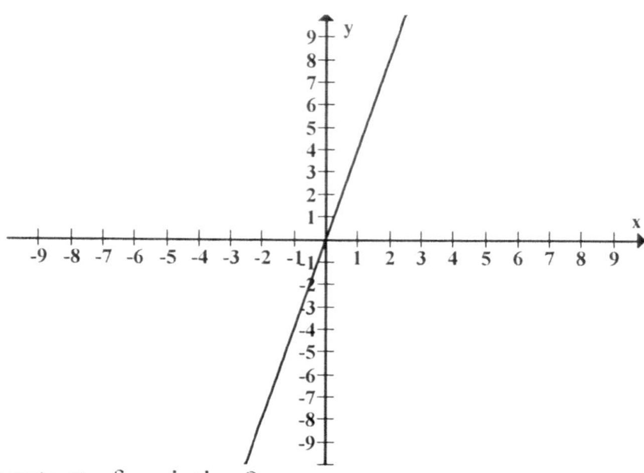

a. What is the constant of variation?

b. Write an equation to describe the variation.

c. Comment on the slope of the graph and the constant of variation value k.

3. The time, t, it takes Girl Scouts to deliver their cookies varies inversely with the number of girls, g. Girl Scouts from Troop 1105 need to deliver 2,600 boxes of cookies. Some of the girls are putting together the orders, while the remaining girls are doing the deliveries. They want to see how many Girl Scouts should deliver cookies to complete the job as quickly as possible. Last year, 40 Girl Scouts were able to deliver 2,600 boxes in 10 hours.

a. Write an equation of variation that gives the time t needed for g Girl Scouts to deliver the boxes of cookies.

b. Find the constant of variation.

c. How long would it take if 32 Girl Scouts were to deliver the same number of boxes?

d. Complete the table:

G	32		50
T		10	

e. Using your equation, estimate the number of Girl Scouts needed to reduce the delivery time to 8.5 hours.

4. The weight of an object on the Moon is directly proportional to its weight on Earth. Jim's weight on Earth is 210 pounds, but he would weigh only 34.65 pounds on the Moon.

a. Find the equation of variation that describes a person's weight, M, on the Moon in terms of his or her weight, E, on Earth.

b. Complete the table below.

E	30	90	150	210
M				

c. Using the table from part (b), approximate the Moon weight of Jim's American Labrador Retriever who weighs 120 pounds on Earth.

d. Jim's friend just adopted a Yorkie Terrier. If the dog weighs only 1.32 pounds on the Moon, what would its Earth weight be?

e. Complete the table to show the average weight of the following breeds of dogs.

Dog breed	Pug	Dachshund	Greyhound	German Shepherd	Mastiff
E	16		67	80	
M		3.3			33

Source: http://www.pgaa.com

5. The registration fees at Valencia Community College for Florida residents during the spring 2008 term were $73.82 per credit hour (excluding admission, lab, or any other special fees). The following table lists the tuition for taking various numbers of credits at Valencia Community College for out-of-state residents.

Credits	3	6	11	12	15
Tuition	$833.01	$1,666.02	$3,054.37	$3,332.04	$4,165.05

Source: http://valenciacc.edu

a. Find the constant of variation.
(Hint: Find the ratio of the tuition to the number of credits.)

b. Express the tuition as a function of the number of credits taken by an out-of-state resident.

c. Graph your function using the x-axis for the number of credits and the y-axis for the tuition value.

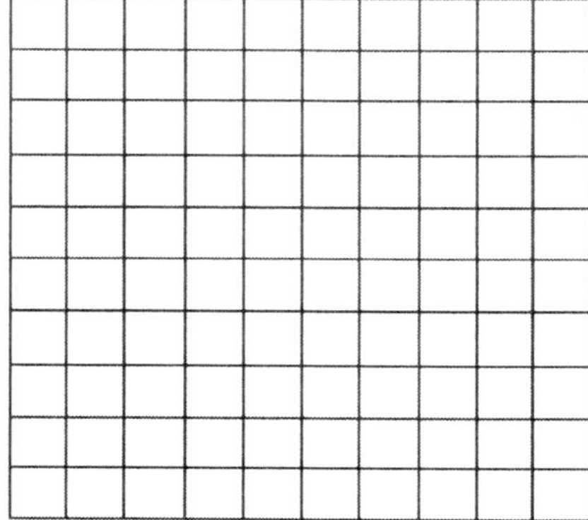

d. Is this function an example of direct variation or inverse variation? Justify your answer.

e. Find $f(11)$ from the given table and corroborate this point with your graph.

f. Estimate the tuition fee for taking 19 credits for an out-of-state resident.

g. Estimate the tuition fee for taking 19 credits for a Florida resident.

h. As of spring 2008, a student has three attempts at Valencia per course, including the original grade, repeat grades, and withdrawals at any point in the term beyond the drop/refund deadline. Upon the third attempt, the student will be charged the full cost of instruction fee, which is equivalent to out-of-state tuition.
Source: http://valenciacc.edu/catalog

What recommendations would you offer a Florida resident student to avoid this penalty?

6 a. Find 3 points on the following graph and complete the table below.

x			
$f(x)$			

b. Use the table to find the constant of variation.

c. Write an equation so that y varies inversely with the square of x.

d. Find the value of x to 3 decimal places for $y = 5$.

e. Without using your calculator, find $f(1.5)$.

f. Use your calculator to find $f(1.5)$ to 3 decimal places. How close is this to your estimate in part (e)?

7. The force, F, needed to loosen a bolt with a wrench is inversely proportional with the length, l, of the handle.

a. Write an equation of variation that describes the force F needed to loosen a bolt in terms of the length of the handle, l.

b. Find the constant of variation, given that 300 pounds of force is needed to loosen a bolt with a 5-inch wrench.

c. How many pounds of force would be needed with an 8-inch wrench?

d. Which wrench would you use if you had to do some repair work on your car without a power tool? Explain your decision.

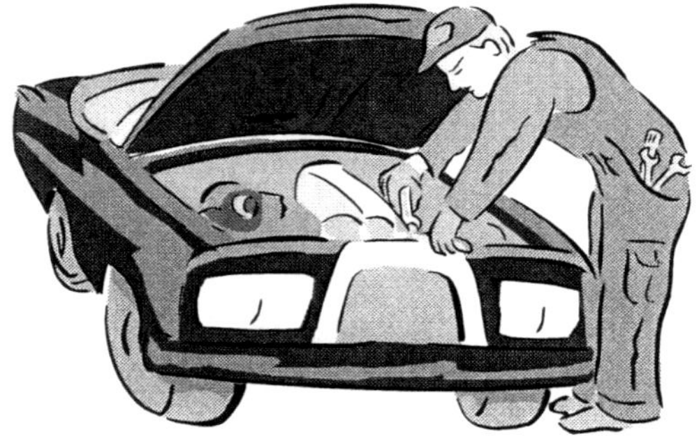

V. Inverse Functions and Applications

Introduction with Numerical and Graphical Approaches

1. Which of the following represents a table of values for the inverse of $f(x) = 3x + 5$? Explain your decision.

a.

x	y
-3	-4
-1	2
4	17

b.

x	y
3	4
1	-5
-4	-17

c.

x	y
-4	-3
2	-1
17	4

Circle one: Table a Table b Table c

Explain: ____

2. Which of the following tables represents a function whose inverse is also a function? Explain your decision.

a.

input	*output*
5	-4
7	5
8	-18
10	12

b.

input	*output*
-5	11
-2	-10
5	11
8	21

Circle one: Table a Table b

Explain: ____

3. Janet prefers to conduct an Internet search for a business phone number instead of using the regular phone book delivered by her telephone service provider. She types the name and location of the business and instantly receives the telephone number. If searching for the phone number via the Internet represents a function, state the inverse of this function, decide whether this inverse would also be a function, and justify your answer.

Inverse: ____

Circle one: Inverse is a function Inverse is a not a function

Explain: ____

4. The graph below shows the relationship between values on the Celsius and Fahrenheit scales. The function displayed, $f(x) = \frac{9}{5}x + 32$, converts a temperature in x degrees Celsius to degrees Fahrenheit. Use the graph of $f(x)$ to answer the following:

a. Estimate $f(10)$ and state its meaning.

b. Estimate $f(30)$ and state its meaning.

The inverse function of $f(x)$ would convert degrees Fahrenheit to degrees Celsius. Use the graph of $f(x)$ to answer the following:

c. Estimate $f^{-1}(32)$ and state its meaning.

d. Estimate $f^{-1}(77)$ and state its meaning.

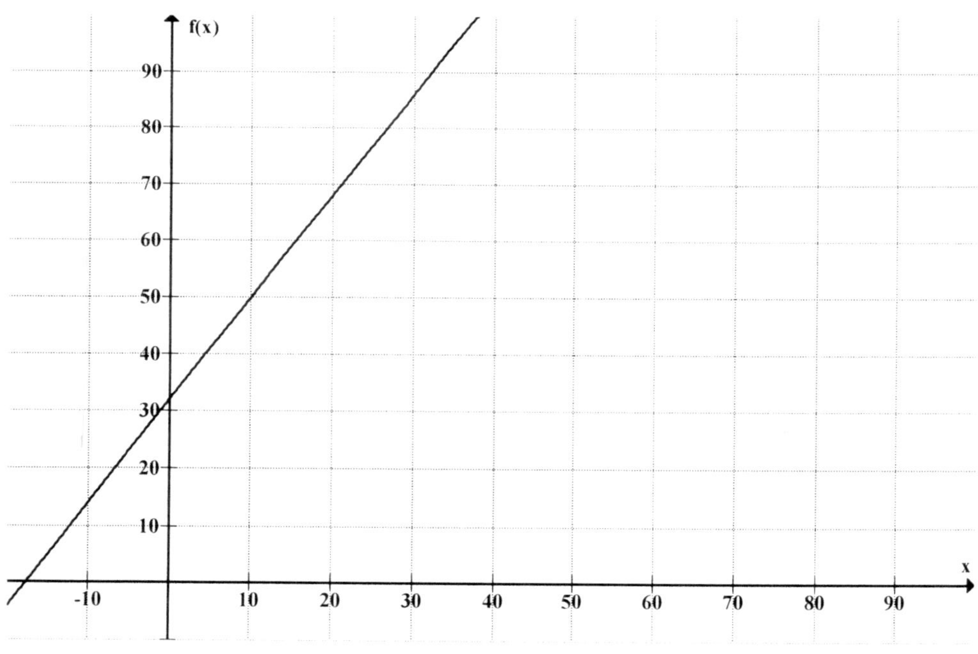

5. Analyze the following figure. Could these graphs represent inverse functions of each other? Justify your decision.

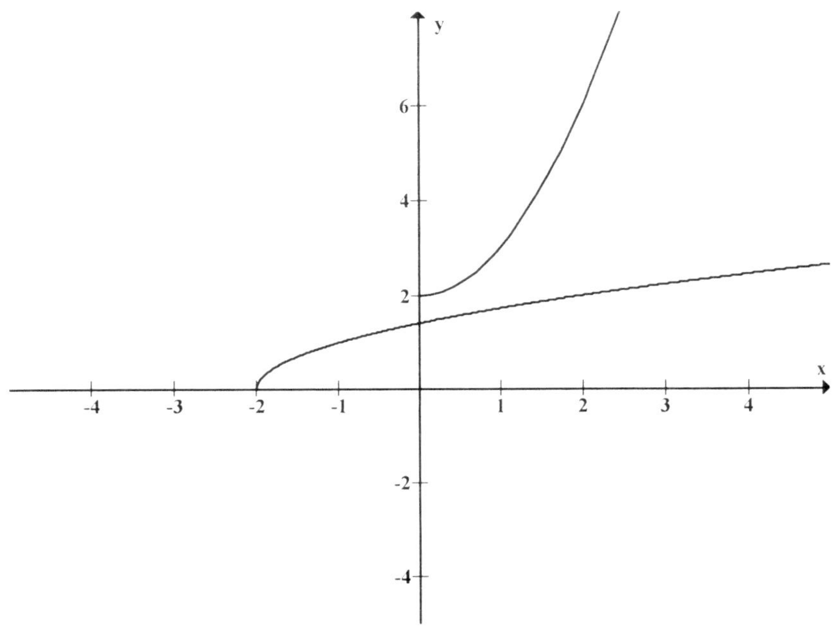

6. The graph of a function, $f(x)$, is shown below. Graph $f^{-1}(x)$.

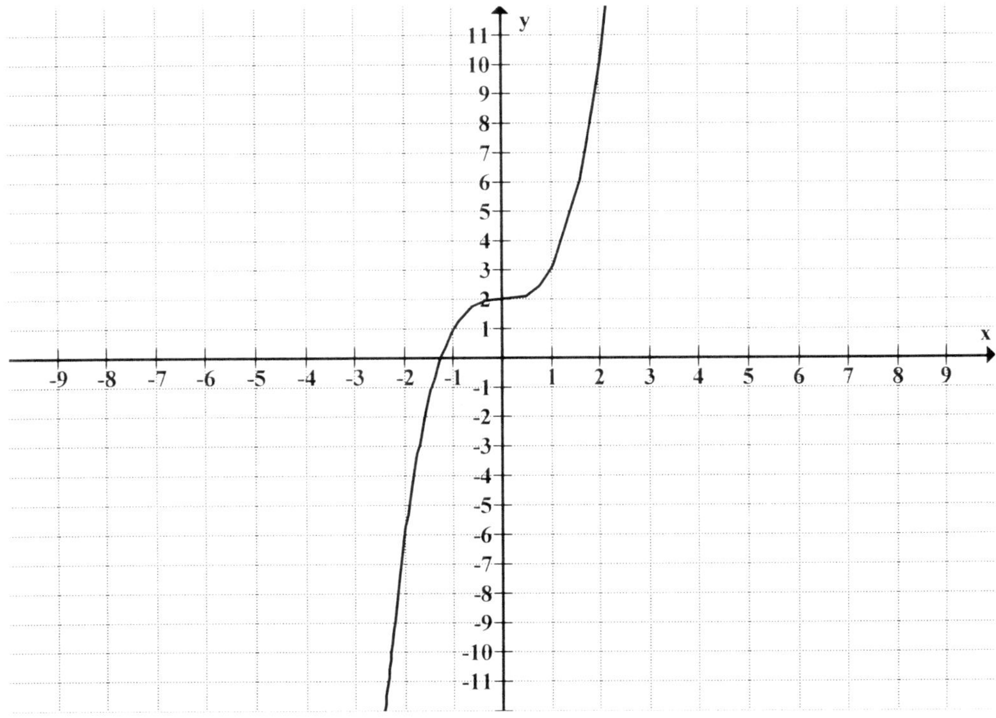

<u>Finding the Inverse Using a Symbolic Approach</u>

1. In a hot air balloon race, the height h of Team Grand Nationals' balloon at t minutes is given by $h(t) = 30 + 50t$.

a. Find $h^{-1}(t)$ symbolically.

b. Find $h(10)$ and state its meaning.

c. Find $h^{-1}(355)$ and state its meaning.

d. Find $h(5)$ and show that $h^{-1}(t)$ undoes the effect of $h(t)$ at $t = 5$. Interpret your answer.

2. Miguel travels frequently by car and has noticed that gasoline prices have increased dramatically in recent years. To save money, Miguel decides to trade his V-8 pickup truck for a 4-cylinder wagon. He will finance his new wagon and has agreed to pay off his $24,222 loan (interest included) in 66 months. His monthly payment will be $367. The function that represents the outstanding amount of money Miguel will pay for this loan is $f(x) = 24{,}222 - 367x$, where x is the number of months.

a. Find $f(60)$ and state its meaning.

b. Find the inverse of the given function symbolically.

c. Find $f^{-1}(11{,}377)$ and state its meaning in the context of this problem.

3. An object is dropped from a height of 100 feet. Its height, h, (in feet) after t seconds is modeled by the function $h(t) = -16t^2 + 100$ for $t \geq 0$.

a. Find $h(1)$ and interpret your answer.

b. Find the inverse of the given function symbolically.

c. Find $h^{-1}(0)$ and interpret your answer.

VI. Quadratic Functions and Modeling

Solving Quadratic Equations Using Different Methods
(Extraction of roots, factoring, quadratic formula, numeric solutions)

Problem 1 deals with compound interest. Here is brief background information: Savings institutions offer accounts on which the interest is compounded annually. Compound interest is interest paid on the principal and also on any interest earned from past years. Over time, compound interest will yield more money than simple interest. The basic formula used to calculate compound interest is $A = P(1 + r)^n$, where A is the final amount of money in the account, P is the principal (initial amount you borrow or deposit), i is the interest rate per year (as a decimal value), and n is the number of investment periods (for this problem, number of years).

1. Chaoxiang deposited $2,964 in a money market account that compounds interest annually.

a. What is the principal for this problem?

b. If Chaoxiang leaves his money in this account for 2 years, replace the known variables on the given formula and rewrite this formula in terms of the interest rate, r.

c. If Chaoxiang wishes to have a minimum of $3,125 in his money market account at the end of the 2-year period, determine the interest rate that will be required. (Hint: Solve your equation by extraction of roots.)

2. A roofer was applying waterproof material in preparation for the hurricane season. He accidentally dropped the roofing nailer from a height of 100 feet. Its height, h, (in feet) after t seconds is modeled by the function $h(t) = -16t^2 + 100$ for $t \geq 0$. (Don't worry, luckily no one was hurt!)

a. Create a table of values for the height formula, starting at 0 seconds and assigning ΔTbl = 0.25. Report your answers for $t = 0.5, 0.75, 1.25, 2, 2.25, 2.5$, and 3. (Check Appendix D if you need a review on creating a table of values with the grapher.)

T	$h(t)$

b. Use your table to determine when the roofing nailer reaches an altitude of 84 feet.

c. How long will it take the nailer to hit ground level?

d. Find $h(2)$ from your table and state its meaning in the context of this problem.

e. If $h(t) = 100$, find t using your table and state its meaning in the context of this problem.

3. Sarah bought a new 50-inch screen plasma HDTV for $2,099.99. A plasma television 50-inch screen is measured diagonally from corner to corner. The width of her new plasma TV screen is 18.3 inches more than its height. Using the quadratic formula, find the measurements of the width and height of the plasma screen.
(Hint: Apply Pythagoras' theorem to help solve this problem.)

Width: ____________________________ Height: ______________________________

4. According to legend, the Italian astronomer Galileo conducted an experiment in 1589 atop the Leaning Tower of Pisa. Galileo dropped two rocks; one was 10 pounds and the other weighed only 1 pound. In his experiment, he theorized that the rocks would hit the ground at the same time. Let $h(t) = -16t^2 + h_0$ denote the height of an object where t is time in seconds and h_0 is the initial height of the object.
Sources: http://pbs.org; http://torre.duomo.pisa.it

a. When would Galileo's rocks hit the ground if the height of the Leaning Tower of Pisa is approximately 180 feet? Will you use factoring, quadratic formula, or extracting roots to solve this problem? Explain your decision.

vorld's tallest building being built as of 2008 is the Burj Dubai in the United Arab
It currently stands at approximately 2,063 feet. Approximate, to 2 decimal
en Galileo's rocks would hit the ground if dropped from this building.
/burjdubai.com

5. Instructor Robertson is getting ready for a Tae Kwon Do exhibition. He is practicing a stunt that involves a front roll with a jump back hook kick.

a. If Instructor Robertson stands on a 55-foot platform, determine when he will safely land on the ground, feet first! Use the height equation $h(t) = -16t^2 + h_0$ to solve this problem.

b. Fellow black belt Instructor García is concerned about the safety of the stunt. He decides to place a 300-square-foot mat that is 3 feet high under the platform. Use the height equation $h(t) = -16t^2 + h_0$ to determine when Robertson will safely land on the mat. (Remember the safety mat is 3 feet high.)

6. For each of the equations below, choose (circle) the appropriate method of solving, explain your selection, and then solve. Estimate to 2 decimal places where needed.

a. $7x^2 - 5x - 1 = 0$ Factoring Extracting roots Quadratic formula

Explain ______________________________

Solution ____________

b. $x^2 - 5x - 6 = 0$ Factoring Extracting roots Quadratic formula

Explain ______________________________

Solution ____________

c. $4x^2 - 11 = 0$ Factoring Extracting roots Quadratic formula

Explain ______________________________

Solution ____________

7. The height in feet reached by a baseball after t seconds is given by the function $h(t) = -16t^2 + 108t$.

a. Use factoring to determine when the ball hits the ground. Write your answer in fraction form, and then convert to a decimal form. Show all work.

Fraction form:

Decimal form:

b. Use factoring to determine when the ball was 182 feet in the air. Write your answers in fraction form, and then convert to a decimal form. Show all work.

Fraction form:

Decimal form:

c. Which of the answers in part (b) represents the ball 182 feet in the air on its way up? Explain your answer.

Parabolas in Standard Form and Characteristics

For problems 1–5, carry out the following by hand:

i. Find the vertex in ordered pair form. Write the answers in fraction form where needed.

ii. Use the vertex to describe the shifting with respect to the standard parabola.

iii. State whether the parabola is concave up or down.

iv. Find the maximum or minimum value.

(Remember that $x_V = -\dfrac{b}{2a}$; use the x_V-value to find the y_V-value)

1. $y = -3x^2 - 12x + 4$

Vertex: Shifting: Concavity: Min. or max:

2. $f(x) = x + 2x^2 - 2$

Vertex: Shifting: Concavity: Min. or max:

3. $y = -5x^2 + 3 + 20x$

Vertex: Shifting: Concavity: Min. or max:

4. $y = 7 - 4x + x^2$

Vertex: Shifting: Concavity: Min. or max:

5. $f(x) = x^2 + 3x - 4$

Vertex: Shifting: Concavity: Min. or max:

Problems 6–8 deal with the nature of the solutions (number of x-intercepts) of a parabola. Here is brief background information:
In the quadratic formula, the quantity under the radical, $b^2 - 4ac$, is called the *discriminant*, and it determines the nature of the solutions of the quadratic equation.

6. If the discriminant is negative —that is, $b^2 - 4ac < 0$ —there are *no real solutions*, and the graph does not cross the x-axis.

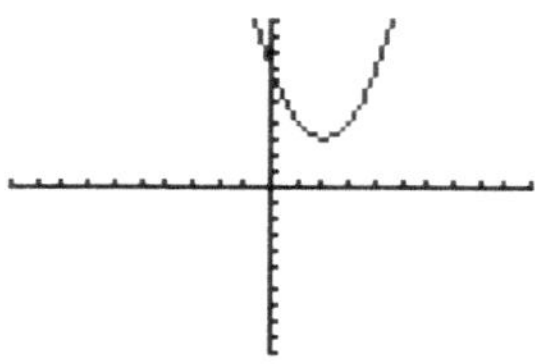

Provide an example of an equation of a parabola that reflects this situation, and demonstrate algebraically that your equation satisfies $b^2 - 4ac < 0$.

7. If the discriminant is zero —that is, $b^2 - 4ac = 0$ —there is *one real solution*, and the graph will touch (not cross) the x-axis only once.

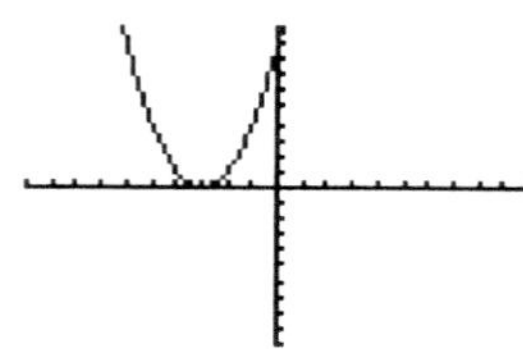

Provide an example of an equation of a parabola that reflects this situation, and demonstrate algebraically that your equation satisfies $b^2 - 4ac = 0$.

8. If the discriminant is positive—that is, $b^2 - 4ac > 0$ —there are *two real solutions*, and the graph will cross the x-axis twice.

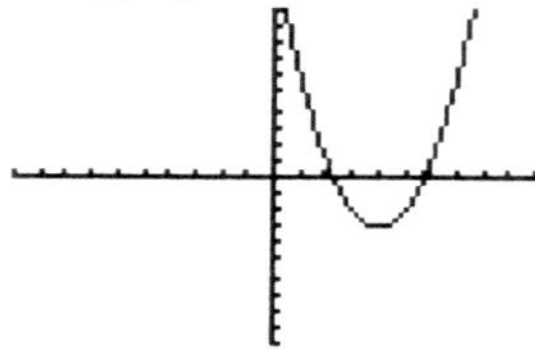

Provide an example of an equation of a parabola that reflects this situation, and demonstrate algebraically that your equation satisfies $b^2 - 4ac > 0$.

9. Find the discriminant for each of the following, and determine the nature of the solutions.

a. $y = 2x^2 + x - 2$

Discriminant: Number of real solutions:

b. $f(x) = -x^2 + 3x - 4$

Discriminant: Number of real solutions:

c. $y = 2x^2 + 4x + 2$

Discriminant: Number of real solutions:

10. Use the graphs below to determine the following:
i. If $a > 0$ or $a < 0$.
ii. If the discriminant is negative, zero, or positive.

a. 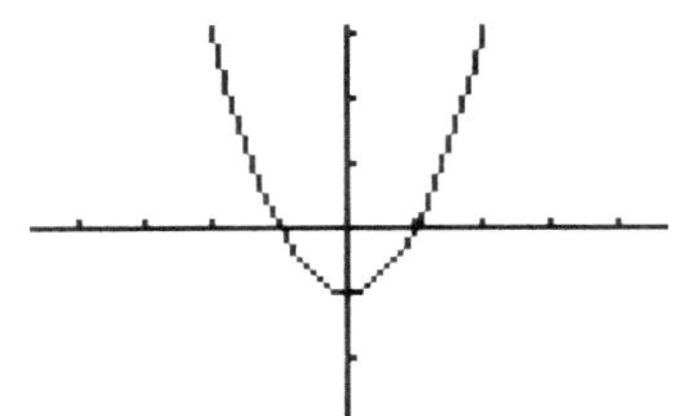

Answer: ________________

b. 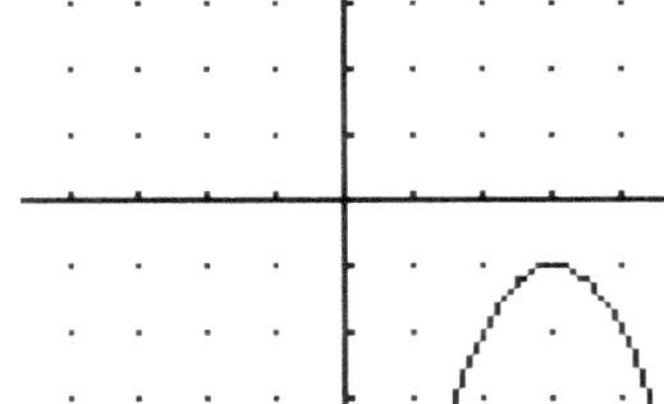

Answer: ________________

c. 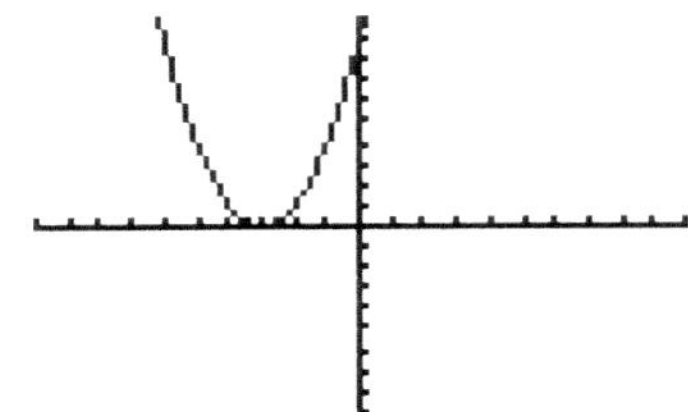

Answer: ________________

d.

Answer: ________________

<u>Graphing Using the Intercepts and the Vertex</u>

For problems 1–2, use the given graph to answer each of the following:

a. Find the x-intercepts.
b. Is $a = 1$ or $a = -1$?
c. Estimate the vertex from the graph.
d. Write the equation of the parabola in standard form using the x-intercepts and the a value.
e. By hand, find the vertex from the standard form in part (d). Write in fraction form if needed.
f. Does your estimate match the vertex found in part (e)?

1.

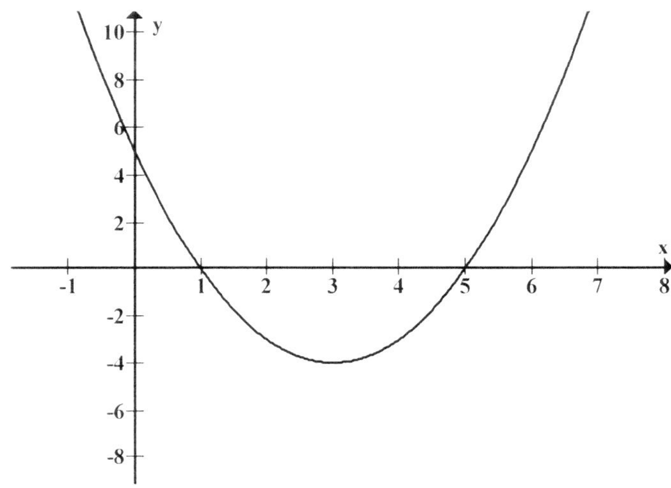

Answers:

a. ______________________

b. ______________________

c. ______________________

d. ______________________

e. ______________________

f. ______________________

2.

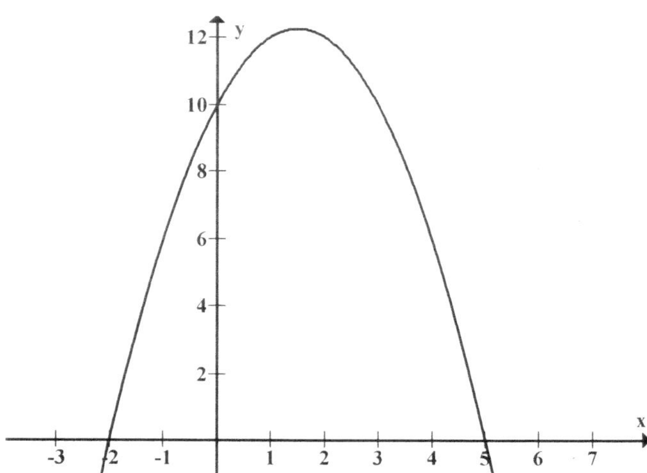

Answers:

a. ______________________________

b. ______________________________

c. ______________________________

d. ______________________________

e. ______________________________

f. ______________________________

3. The Johnsons' subdivision celebrated New Year's Eve with a big neighborhood party. As part of the festivities, the family watched a fireworks display at midnight. The fireworks were fired vertically into the air from the ground level at an initial velocity of 80 feet per second. The following function represents the height of the fireworks, h, at t seconds:

$h(t) = -16t^2 + 80t$

a. Find the coordinates of the vertex of this quadratic function algebraically.

b. Explain the significance of the vertex in the context of this problem.

c. Find the x- and y-intercepts of the height function.

d. Explain the significance of the intercepts in the context of this problem.

e. Use the vertex and intercepts to graph this quadratic function by hand. Label your axes and show your scale.

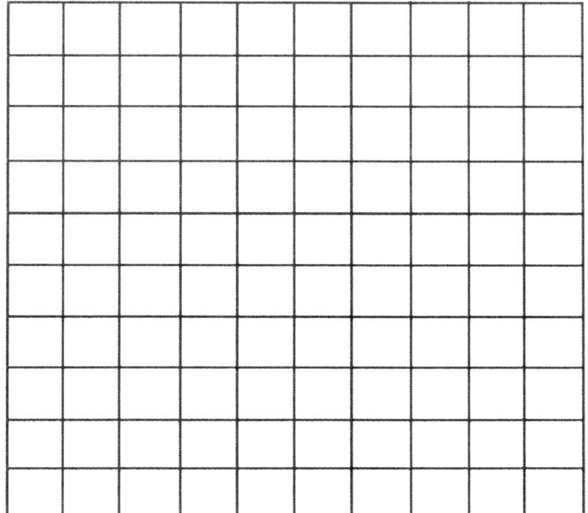

Parabolas in Vertex Form

1. The following parabolas are written in vertex form. Graph each of these quadratic functions with your graphing calculator and explain how the changes in a, h, and k transform the graph of each function when compared to the graph of $y = x^2$.
Note: (h, k) represent the coordinates of the vertex.

a. $y = (x - 4)^2 + 2$
b. $y = 3(x - 4)^2 + 2$
c. $y = -3(x - 4)^2 - 1$
d. $y = -\frac{1}{2}(x + 4)^2 - 4$

Show a rough sketch of your graphs here:

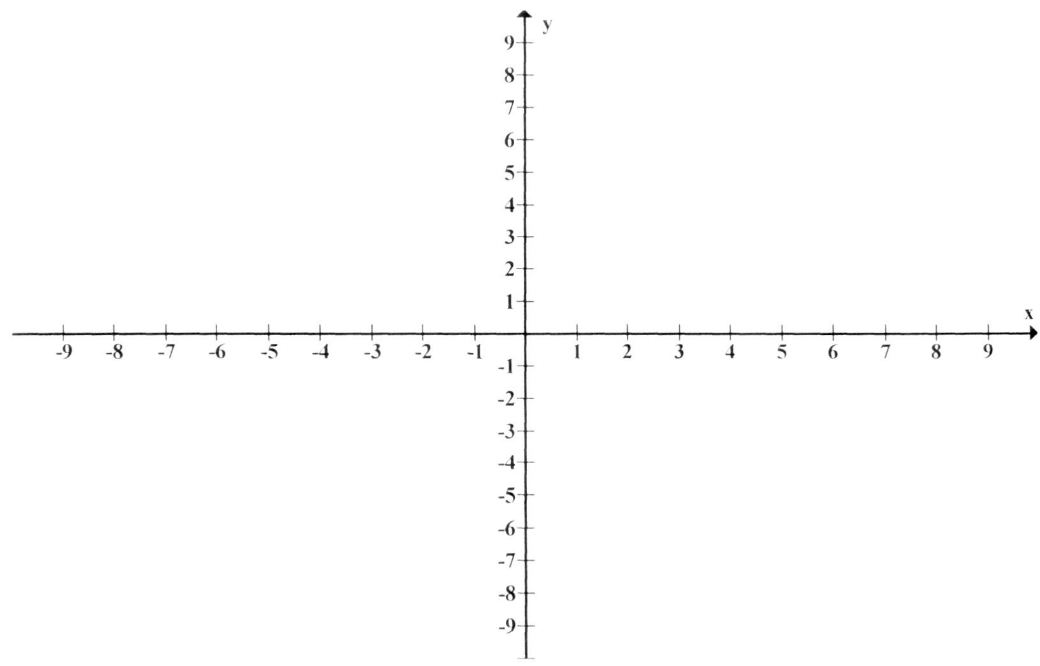

Explain: ____

____

____

2. Given the following graphs, write the vertex form of each parabola. Assume that $a = 1$ or $a = -1$. Assume each tick mark is 1 unit

a.

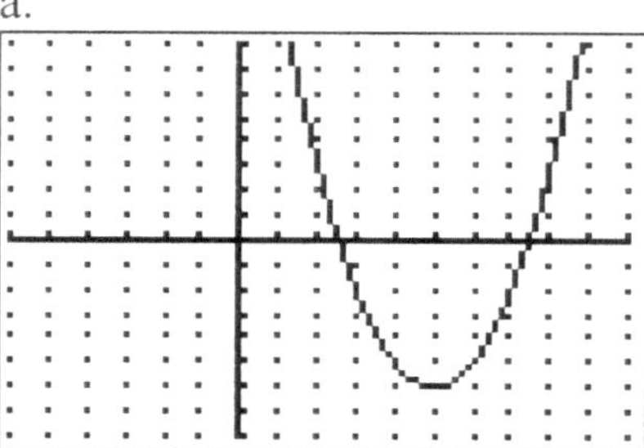

Equation: _______________

b.

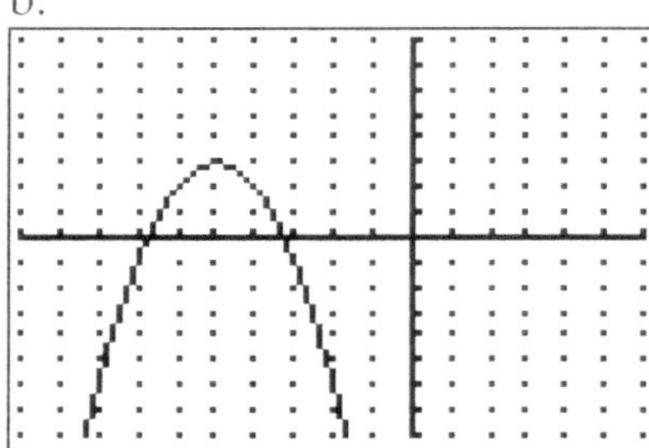

Equation: _______________

c.

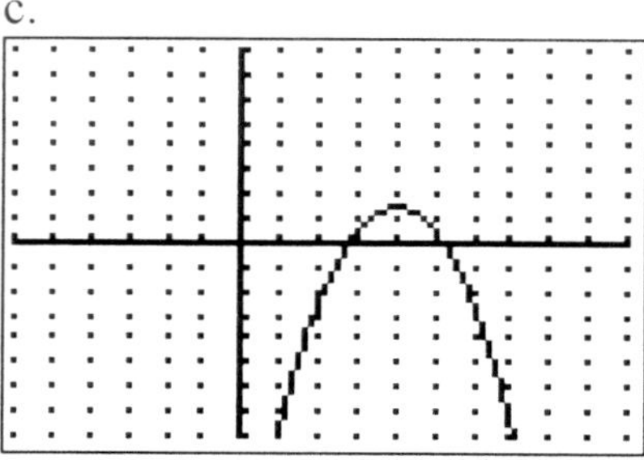

Equation: _______________

d.

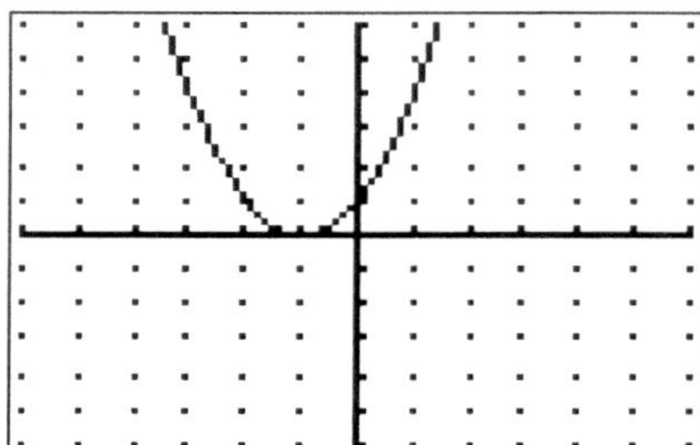

Equation: _______________

3. The number of fatal motor vehicle accidents in the United States from 1990 to 2005 can be approximated by the function $f(x) = 34(x - 8)^2 + 37{,}615$, where x is the number of years after 1990.

Source: http://www.census.gov

a. Find the vertex of this quadratic model and interpret its meaning in the context of the problem.

b. Find $f(10)$ and state its meaning.

c. If $f(x) = 39{,}791$, find x and state the significance of your findings in terms of this problem.

4. Construction on the Golden Gate Bridge in San Francisco, CA, began in January 1933 and was completed in April 1937. The height of each tower above the roadway is 500 feet, the distance between the two towers is 4,200 feet, and the bridge itself is 220 feet above water level. The fixed suspension cables from one tower to the other form a parabolic shape.
Source: http://goldengatebridge.org

a. Label and give the coordinates of the vertex, and explain its meaning.

b. Label and give the coordinates that denote the top of the towers.

c. Use the vertex and one of the coordinates from part (b) to find the a value for the vertex form $y = a(x - h)^2 + k$, where (h, k) represent the coordinates of the vertex. Then write the equation in standard form.

Answer for part(c) ____________________________

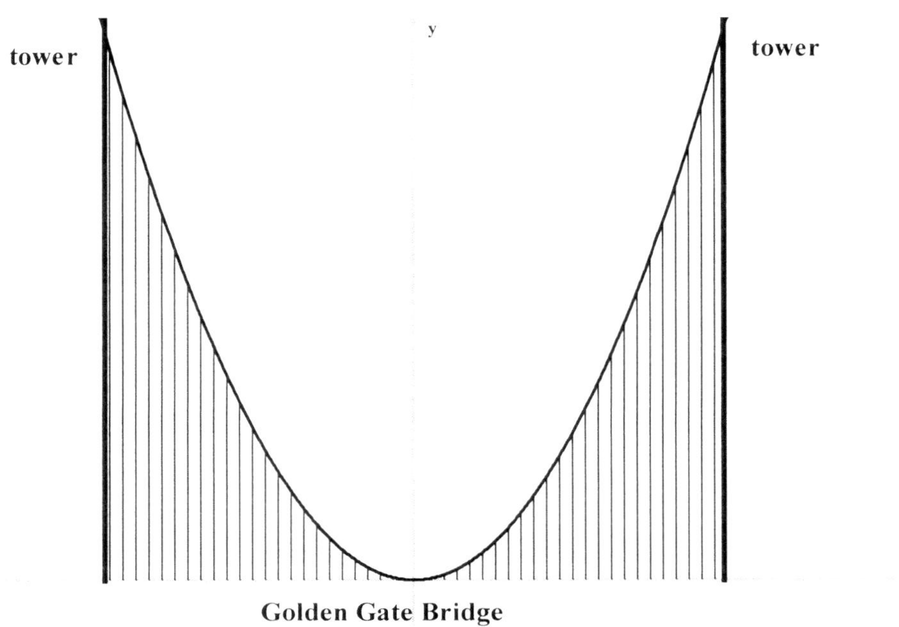

Maximum and Minimum/Break-Even

1. Eigram and Alesig wrote a best-selling book on dog training. The number of copies sold in the first 16 weeks after the book was released can be approximated by the function $f(x) = -203x^2 + 3{,}381x + 1{,}752$, where x is the number of weeks.

a. Find the coordinates of the vertex algebraically. Round your answer to the nearest integer.

b. Without graphing the function, determine whether the vertex is a maximum or a minimum. Justify your answer.

c. Interpret the meaning of the vertex in the context of this problem.

d. Use your graphing calculator to graph the function and include a rough sketch on separate paper. Label your axes and state the chosen viewing window.
(Hint: Use the coordinates of the vertex and the vertical intercept of the function to help find an appropriate viewing window.)

e. Use your graph to determine the week where the number of copies sold would fall below 10,000. (Hint: Graph the line $y = 10{,}000$, and use the intersection method to help in your calculation.)

2. The ancient Colosseum in Rome, Italy, was completed in the year 80 AD. By coincidence, there are 80 arches on every floor. The arches on the bottom floor can be modeled by the equation, $f(x) = -x^2 + 4.2x + 3.09$, where x denotes the width and $f(x)$ denotes the height in meters.
Source: http://thecolosseum.net

a. Find the maximum value.

b. Explain the meaning of the coordinates found in part (a).

c. Find the width of the base of the arch. (Hint: Apply what you know about the axis of symmetry.)

3. The Eiffel Tower in Paris, France, was completed in 1889. Gustave Eiffel designed the tower and its construction took approximately 2 years and 2 months, during which Stephen Sauvestre was the main architect. The shape of the tower's base (inside measurements) can be approximated by $f(x) = -x^2 + 74.24x - 1338.49$, where x denotes the width and $f(x)$ denotes the height in meters.
Source: http://tour-eiffel.fr

a. Comment on the shape of the base of the tower.

b. Find the maximum height of the arch and explain its meaning.

c. Francois is standing on the top of the base of the arch. If he decides to throw his replica of the Eiffel Tower to his friend Jean-Claude, who is standing on the ground, how long will it take for his replica to reach the ground? Let $h(t) = -16t^2 + 39.4$ denote the height in meters at t seconds. State which method you used to answer the question.

Problems 4–6 deal with the concepts of revenue, cost, and profit. Here is brief background information:
Recall that Profit equals Revenue minus Cost; that is, $P(x) = R(x) - C(x)$. If a company's total revenue equals its total costs — that is, $R(x) = C(x)$ — it is said that the company *breaks even*. In other words, the profit equals 0.

4. The profit of a jeans manufacturer that specializes in destroyed jeans, embroidered jeans, and painted jeans is modeled by the function $P(x) = -0.10x^2 + 60x - 3500$, where x represents the quantity of jeans sold.

a. How many jeans would the manufacturer need to sell to break even? Round to the nearest whole number, if needed. Solve algebraically or graphically.

b. Determine the number of jeans the manufacturer would need to sell to obtain a maximum profit, and state the maximum profit. Solve algebraically or graphically.

5. Karen is making pendants to raise money for breast cancer awareness month. She estimates that her fixed costs will be \$440, and the cost to make each pendant will be \$10. Karen decides to consult her friend Charlotte, a math whiz, to determine the approximate revenue she will generate for the month. Charlotte thinks the price for each pendant should be \$27 but suggests discounting the price by 10% due to the economy. Therefore, Charlotte comes up with the price function, $P(x) = 27 - 0.10x$.

a. Using the graph below, approximately how many pendants would Karen have to sell to break even?

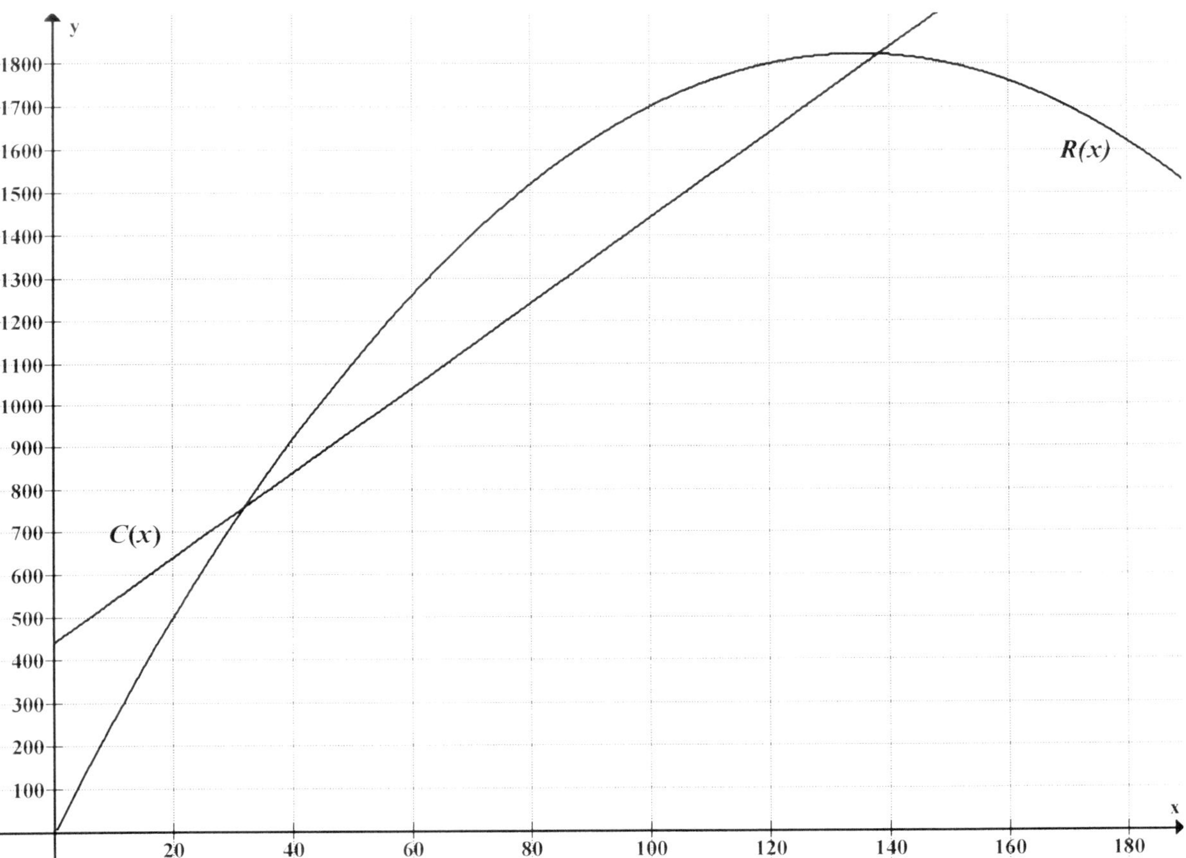

b. Write an equation that represents the revenue function $R(x)$.

c. Write an equation that represents the cost function $C(x)$.

d. Find the approximate cost and revenue at the break-even points.

e. Write an equation that represents the profit function $P(x)$.

f. Using the profit function obtained in part (e), solve algebraically to determine the break-even points.

g. How do your estimates found in part (a) compare with the values found in part (f)?

h. Solve algebraically to determine the number of pendants Karen would need to sell to obtain a maximum profit, and state the maximum profit.

6. A manufacturing company's revenue can be denoted by $R(x) = 150x - 0.15x^2$, and its costs can be modeled by $C(x) = 11{,}975 + 25.50x$.

a. Using your grapher, let $y_1 = C(x)$ and $y_2 = R(x)$, to find the break-even point on the *left* side of the graph. Use the following window: *X*min = -220, *X*max = 900, *X*scl = 100, *Y*min = -7,875, *Y*max = 37,500 and *Y*scl = 20,000. Include a rough sketch of your graph on separate paper.

b. Write the break-even point found in part (a) as an ordered pair, and interpret its meaning.

c. Complete the table below by setting the TblStart = 715 and the ΔTbl = 1.

x						
y_1						
y_2						

d. Using the table from part (c), approximate the other break-even point.

e. Use your grapher to verify the break-even point; write the point in ordered pair form.

VII. Exponential/Logarithmic Functions and Applications

Exponential Functions and Their Graphs

1. Assume that the following situations represent exponential functions. Determine whether the functions are increasing or decreasing.

a. Fish population with increased waste runoff in the lake

b. Price of gas when demand exceeds supply

c. The number of *American Idol* viewers as it nears the season finale

d. Amount of time a typical college student studies during Spring Break

e. Gallons of bottled water sold during an active hurricane season

Answers:

a. ________________________________

b. ________________________________

c. ________________________________

d. ________________________________

e. ________________________________

2. Using a complete sentence for each, give a situation that would denote an increasing exponential function and a decreasing exponential function.

a. Increasing exponential function ____

____

b. Decreasing exponential function ____

____

3. Match the following exponential functions with their graphs. Fill in each blank with the correct graph: y_1, y_2, y_3, or y_4.

a. ______ $= 0.25^x$ b. ______ $= 2^x$ c. ______ $= \left(\frac{1}{2}\right)^x$ d. ______ $= 3^x$

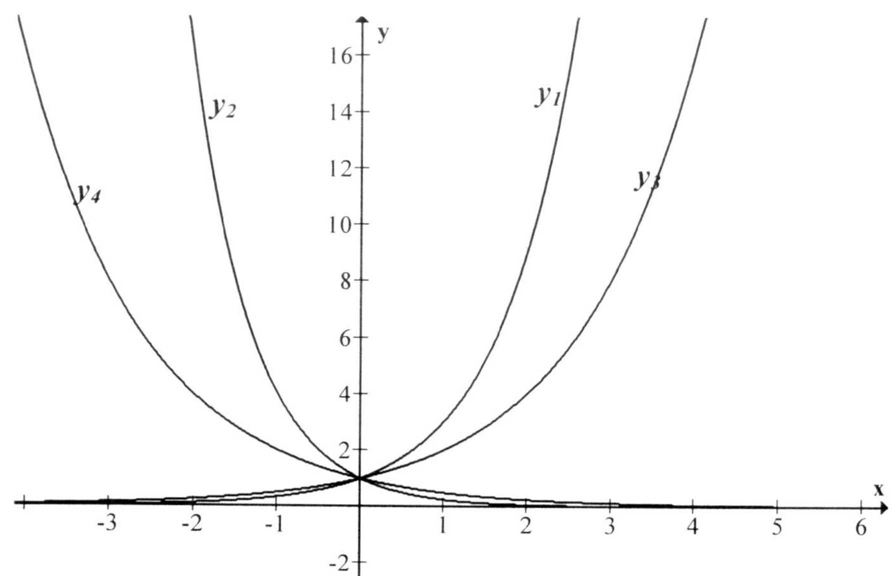

4. Using the graph of $y = 3^x$ as a reference, describe the change to the graph given the following equations:

Answers:

a. $y = -3^x$ a. ______________________________

b. $y = 3^{-x}$ b. ______________________________

c. $y = 3(3)^x$ c. ______________________________

d. $y = 3^x - 2$ d. ______________________________

e. $y = 3^{x-2}$ e. ______________________________

f. $y = \frac{1}{2}(3)^x$ f. ______________________________

5. Use the graph below to answer the following:

a. Find the y-intercept.

b. Choose four ordered pairs to complete the table. Write all values in fraction form.

x				
$f(x)$				

c. Using the table above and the formula $y = a^x$, write an exponential function that describes the graph.

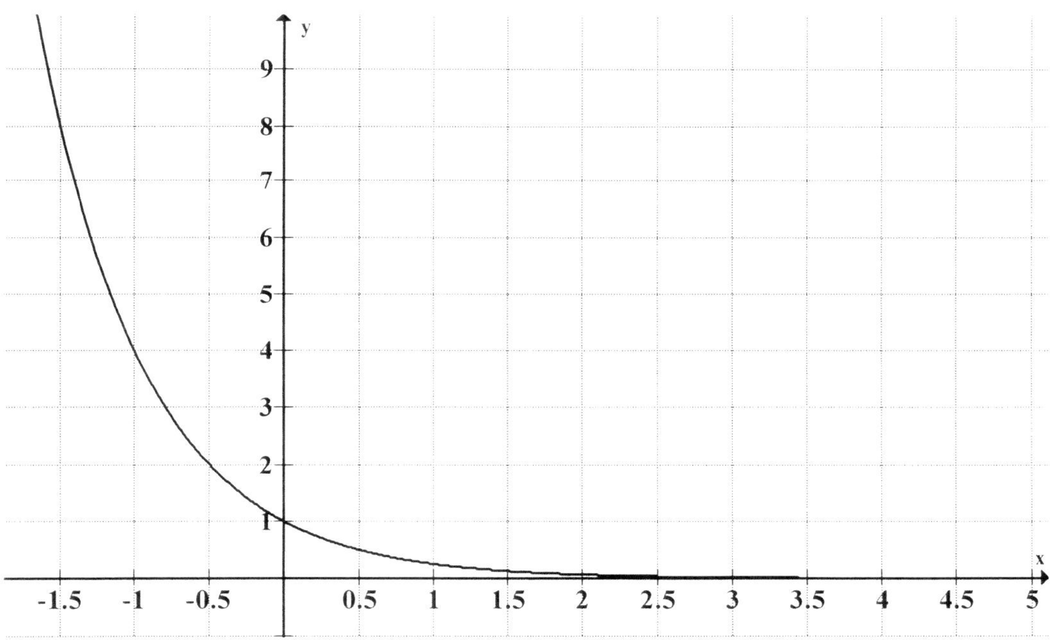

Answer:__________________________

Logarithmic Functions and Their Graphs

1. Use the function $f(x) = \log(x)$ to answer the following.

a. Write a function that increases more rapidly. $f(x) =$ ____________

b. Write a function that is shifted to the right 4 units. $f(x) =$ ____________

c. Write a function that is shifted down 2 units. $f(x) =$ ____________

d. Write a function that is shifted up 5 units and shifted left 1 unit. $f(x) =$ ____________

2. Using the graph of $y_1 = \log(x)$ as reference, write an equation for the graphs of y_2 and y_3.

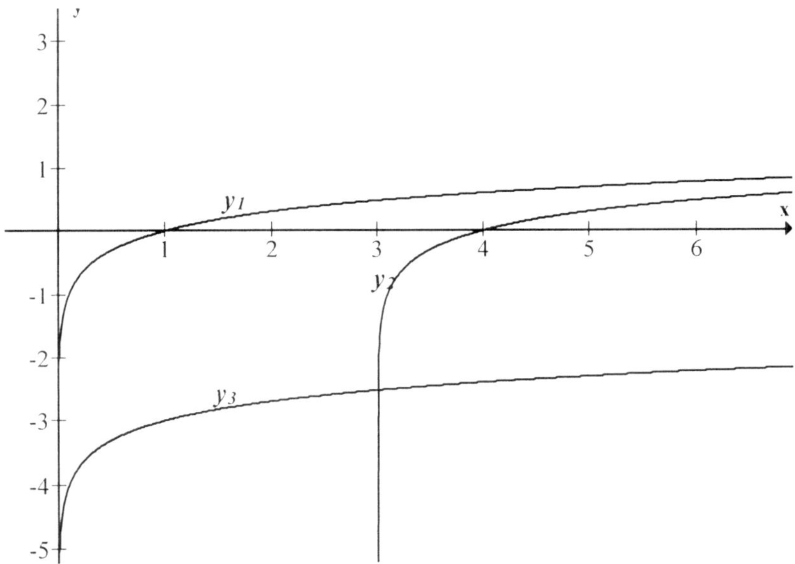

Answers:

$y_2 =$ ______________________________

$y_3 =$ ______________________________

3. Given $y = \log(x)$, answer the following.

a. Complete the table.

x	1	10	100
$y = \log(x)$			

b. State the domain and range of $y = \log(x)$ for the table values in part (a).

Domain ______________________ Range ______________________

c. Find the inverse of $y = \log(x)$ and write it within the table; complete the table of values.

x	0	1	2
Inverse =			

d. State the domain and range for the inverse of $y = \log(x)$ for the table values in part (c).

Domain ______________________ Range ______________________

e. Comment about the domain and range values for the two functions.

4. Use the function $y = 4^x$ to answer the following.

a. Complete the table of values, then graph the function.

x	- 1	0	1	2	3
$y = 4^x$					

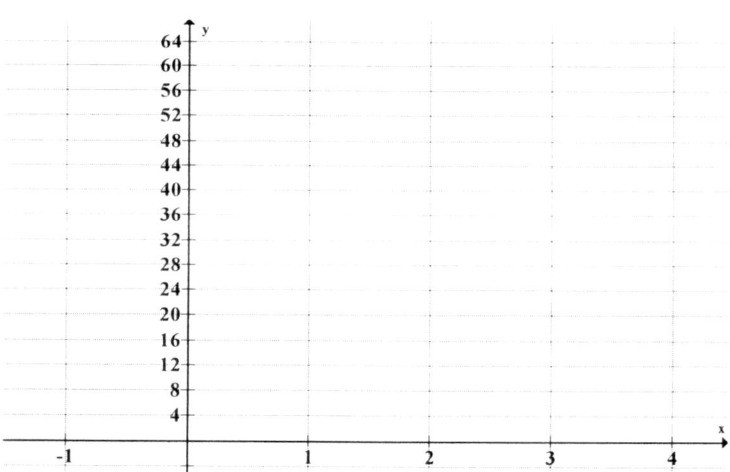

b. Using the table of values in part (a) and your knowledge of inverses, complete the table for the function $y = \log_4(x)$.

x					
$y = \log_4(x)$					

c. Graph the function $y = \log_4(x)$ and label all tick marks.
Hint: Use the tick mark labeling in part (a) as a guide.

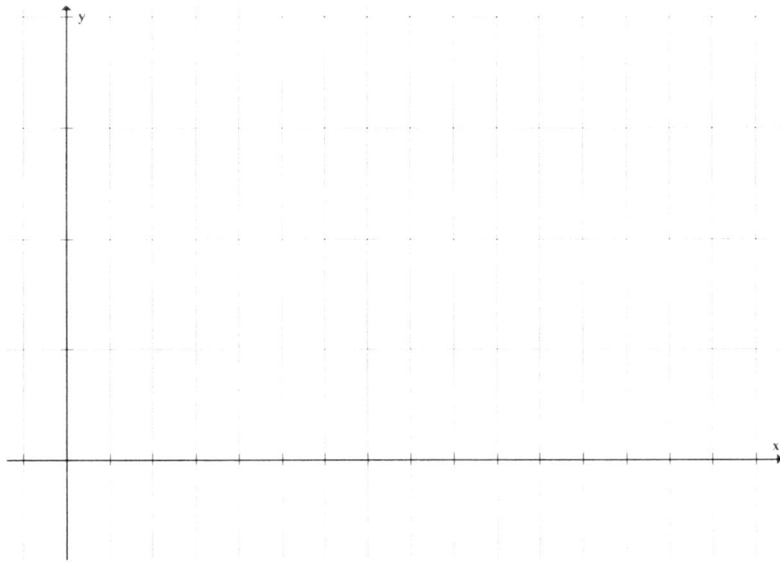

5. a. Graph the function $y = \ln(x)$. Label four points on the graph, including any x or y intercepts.

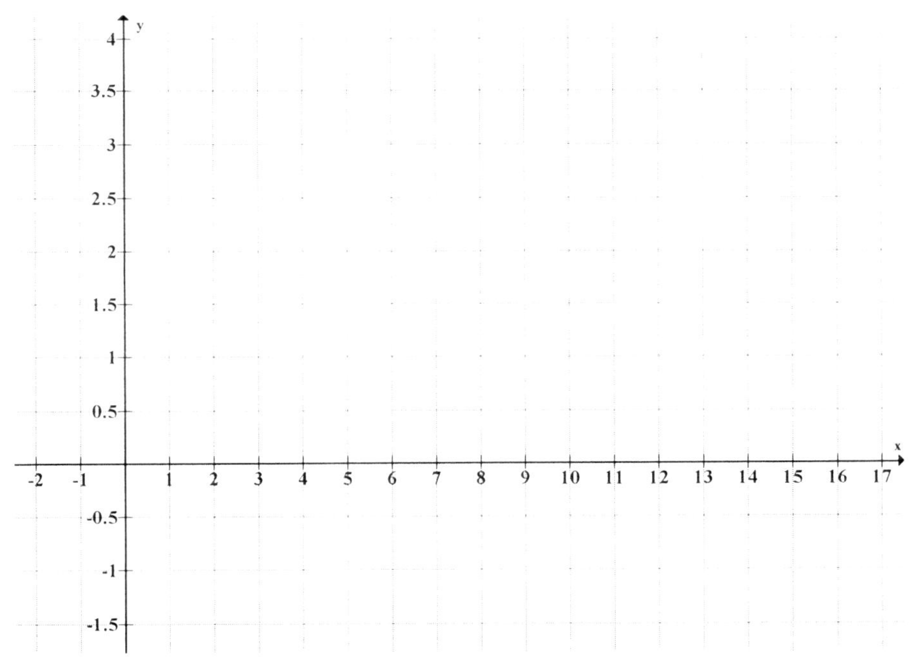

b. Graph the function $y = e^x$. Label four points on the graph, including any x or y intercepts.

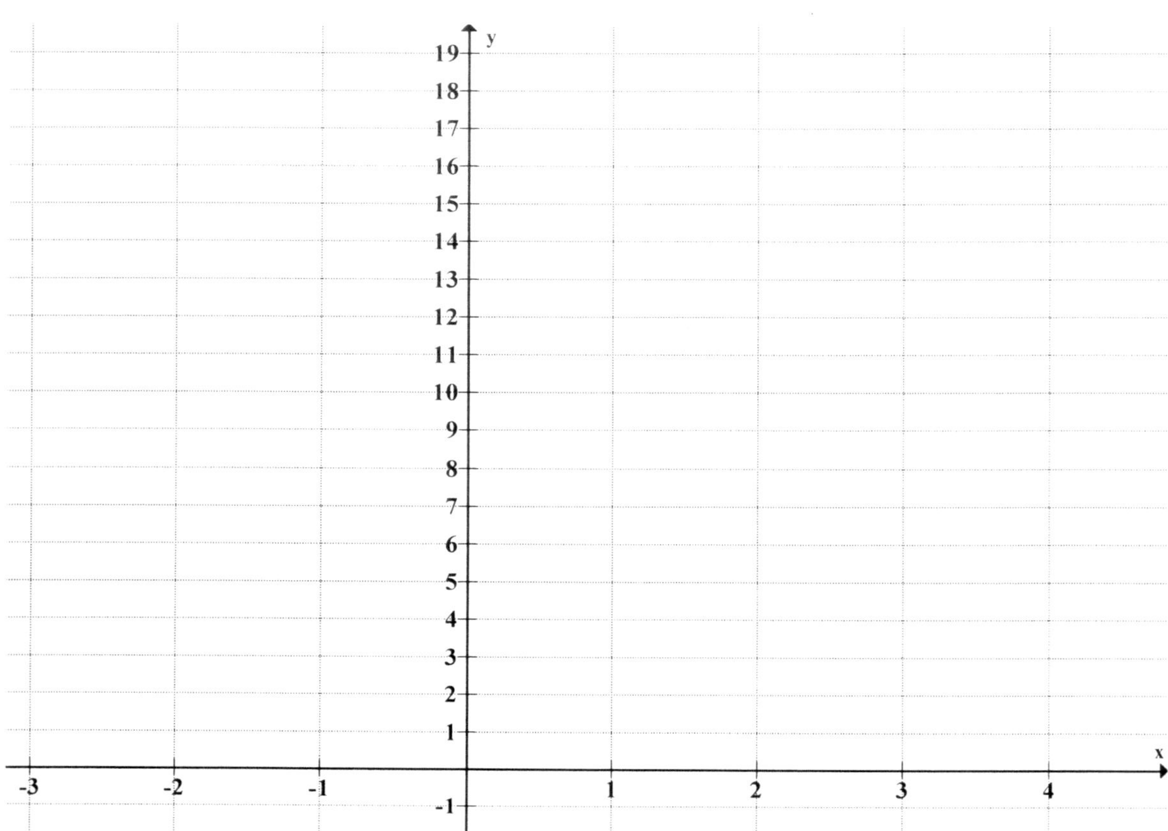

c. Complete the table of characteristics for the functions $y = \ln(x)$ and $y = e^x$.

	$y = \ln(x)$	$y = e^x$
x-intercept		
y-intercept		
domain		
range		
asymptote		

Solving Exponential and Logarithmic Equations

1. Apply the same-base property to solve each equation algebraically.

a. $6^{x-2} = \sqrt{6}$ x = ________

b. $5^{x+4} = \frac{1}{125}$ x = ________

c. $256 = 4^{x^2 - 4x - 8}$ x = ________

2. Use the equations from problem (1) to answer the following.

a. Explain how we can solve each equation *graphically*.

b. Solve the equations from problem (1) graphically. For each case, include a rough sketch of your graph and state your viewing window.

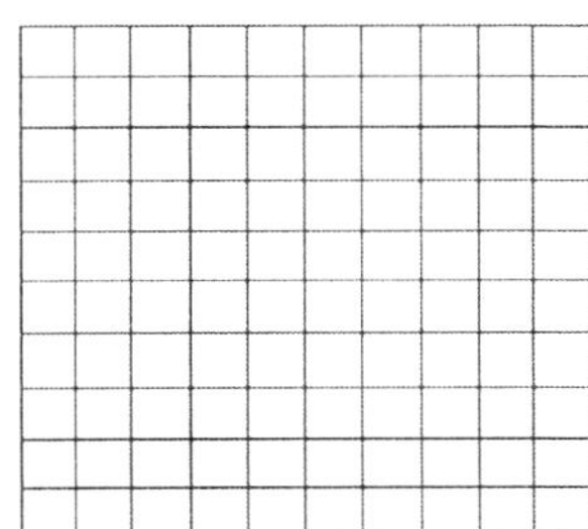 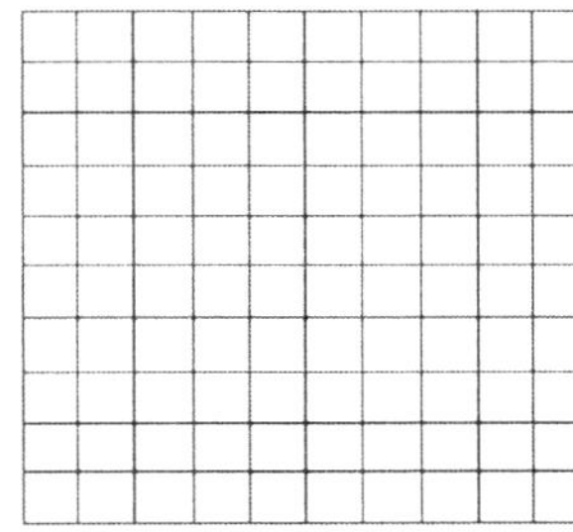 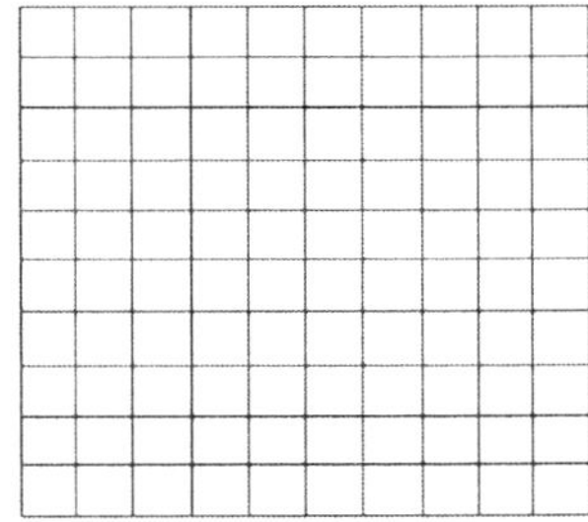

c. Compare your answers with those for problem (1).

d. Which method of solving did you prefer and why?

3. Determine if applying the same-base property is an appropriate method to solve the exponential equation $7^{-x} = 25$. Justify your decision.

4. Approximate the solution to $7^{-x} = 25$ graphically and state your answer to the nearest thousandth.

5. Frederick solved the equation $8^{x+5} = 2^x$ algebraically. His work is shown below.

$$8^{x+5} = 2^x$$
$$(2^3)^{x+5} = 2^x$$
$$2^{3x+5} = 2^x$$
$$3x + 5 = x$$
$$2x = -5$$
$$x = -5/2$$

a. His instructor marked it wrong. Why?

b. Correct the calculations and provide the answer.

6. Juanita and Pivo are in the tutoring center working on their math homework. Juanita is having trouble solving the equation $3 + 2x = 6$. Coincidentally, Pivo is working on the equation $3 + 2\ln(x) = 6$. Professor Epstein notices her students are having trouble, so she tells Pivo to solve her equation using the same steps as Juanita would to solve her problem but advises her not to forget one important additional step.

a. Using complete sentences, explain the steps Juanita would use to solve her problem.

b. Using a complete sentence, explain the extra step Pivo would take to solve her equation.

c. Solve Pivo's equation and write the answer in exact form, then round to 3 decimal places.

Exact answer ________________ Answer to 3 decimal places ________________

d. Complete the table below to verify your answer. Set the TblStart = 3 and ΔTbl = 0.5

x	y_1	y_2

7. Use the equations $\log(x) + \log(x - 48) = 2$ to answer the following.

a. Find the solution(s) algebraically; show all steps.

b. Are both of your x-values from part (a) solutions to the given equation? Why or why not?

c. Solve the equation graphically by using the intersection method. Use the window [-10, 200, 10] by [-2, 6, 1], and then make a rough sketch below.

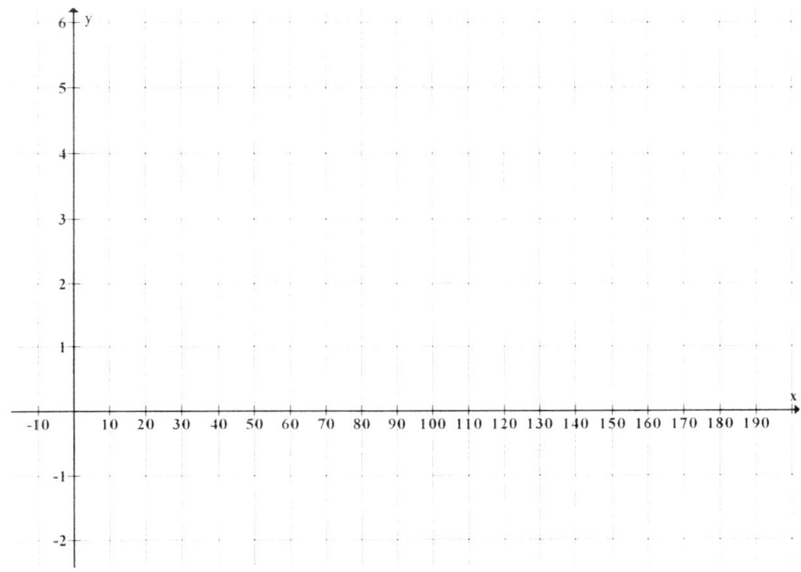

d. How many solutions does the graph show? Explain.

8. Solve the equations by hand, show all work, and give only exact answers.

a. $3\log_2(x) = 4$

b. $\log_2(3x) = 4$

c. $\log_2(3x + 1) - \log_2(x) = 4$

d. $3 + 2\ln(x) = 4$

e. $\ln(x^3) - 3 = 4$

f. $3e^{3x} - 2 = 4$

9. Naomi is trying to solve $3e^x = 4$. Her answer is $x = \frac{\ln(4)}{3}$. Is she correct? Justify your decision.

10. Refer to the equation $5^{x+1} = 7^x$ for the following questions. State all answer to the nearest thousandth.

a. Solve the equation by taking the *common* logarithm of both sides. Show all your work.

b. Solve the equation by taking the *natural* logarithm of both sides. Show all your work.

c. Solve the equation *graphically*. Show a rough sketch of your graph.

d. Based upon your answers to questions (a) through (c), what can you conclude about the solution method to an exponential equation?

History note: John Napier (1550–1617) was the inventor of logarithms. He was a Scottish mathematician who also introduced the decimal point in writing numbers.

<u>Growth and Decay Models</u>

1. Nigel's father gave him a basketball rookie card from 2005 that is worth $5. The rookie demonstrates potential to become a good NBA player.

a. If the value of this card is expected to have a 12% yearly growth rate, how many years will it take until the card is worth $150? Answer to the nearest year.

b. If the growth remains the same, determine the rookie card's value in 50 years. Answer to the nearest dollar.

c. Assume that another rookie comes along and shows a much higher potential, thus devaluating Nigel's card. In 3 years, his card will only be worth $1.08. Determine the decay factor. Answer to the nearest tenth.

2. Patrick bought a computer for $1,800 for his new dog boarding business. For tax purposes, he estimates that it will depreciate in a way that each year the computer will be worth two-thirds of its value for the previous year.

a. What is the decay factor for this problem?

b. If $V(x)$ represents the current value of Patrick's computer for x years, write an exponential function that models this situation.

c. Use your function to determine the computer's value in 3 years. Answer to the nearest dollar.

d. Graph the function with your calculator for a timeframe of 6 years. Include a rough sketch of your graph; label the axes and show your viewing window.

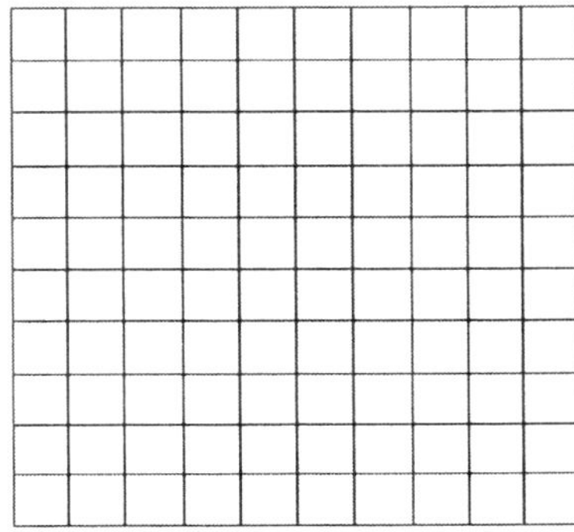

e. Use your graph to estimate the number of years that will pass until Patrick's computer is worth \$350. Answer to the nearest year.

3. Judy's doctor prescribed a medication for her pet allergy. As time passes, the medicine's effectiveness will diminish; for each hour, the drug is only 90% as effective as the previous hour. The doctor prescribed Judy a dosage of 50 milligrams three times a day.

a. If $M(t)$ represents the amount of medicine at t hours, write an exponential function that models this situation.

b. If the initial dose was 50 milligrams and Judy took the medicine 8 hours ago, determine after how long the initial dose will drop to a quantity of 30 milligrams. Solve algebraically and answer to the nearest hour.
(Hint: Take the log of both sides and apply the properties of logs to solve your equation.)

c. Graph $M(t)$ with your calculator and confirm your answer from part (b). Include a rough sketch of your graph; label the axes and show your viewing window.

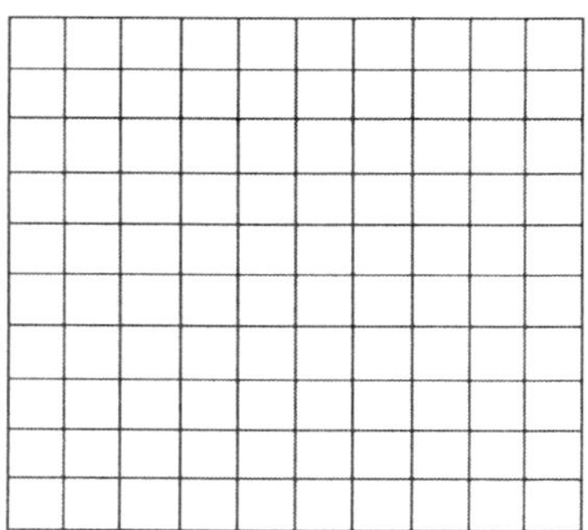

d. Use your graph from part (c) to answer the following question: If Judy only takes her medication one time, what is the amount of medicine left in her system to combat her pet allergy? Round your answer to 3 decimal places.

4. Walt Disney World in Florida is located near Osceola County. The county experienced tremendous growth between the years 2000 and 2006. In 2000, the population was 174,200 and in 2006 the population grew to 244,045.The following table illustrates the county's growth for the years 2000–2006.
Source: http://www.census.gov

Total Population	Osceola County, Florida
July 1, 2006	244,045
July 1, 2005	231,482
July 1, 2004	220,217
July 1, 2003	206,095
July 1, 2002	194,281
July 1, 2001	183,443
July 1, 2000	174,200

a. Using $P(t) = P_0 b^t$ and the information from the table, identify $P(6)$, P_0, and t, and explain their meaning.

b. Write the exponential growth equation that approximately models Osceola County's population for the year 2006.

c. Find the growth factor to 2 decimal places.

d. Find the percent rate of growth and interpret its meaning.

e. When will the population of Osceola County reach 350,000, if the current growth rate continues? Round to the nearest year and give the calendar year as well.

5. Instructor Tanya opened an after-school martial arts class. She started with 10 students and finds that her class increases by a factor of 1.5 students every 2 months. Use the function $P(t) = P_0 b^t$ to answer the following.

a. Identify the initial population and the growth factor.

b. Write the exponential growth function that models this problem.

c. How many students will she have after 5 months? Round to nearest whole number.

d. How many students will she have after 1 year? Round to nearest whole number.

6. Identify each situation as linear or exponential growth.

a. Consumer prices have been rising with inflation at 3.9% per year.

Circle one: Linear growth Exponential growth

b. The freshman student population at a college is increasing about 6.2% per year.

Circle one: Linear growth Exponential growth

c. A boy saves $5 of his allowance every week to buy a new video game.

Circle one: Linear growth Exponential growth

d. A new type of bacteria is quadrupling in population every 8.5 hours.

Circle one: Linear growth Exponential growth

e. An athlete's bowling average has increased by 10 points every year.

Circle one: Linear growth Exponential growth

7. You are a student who will graduate summa cum laude from college next summer and will be offered a job at a prestigious and impressively high-paying corporation. Your new supervisor plans to test your mathematics knowledge, critical and logical thinking skills, and ability to perform mental calculations. She will let you choose your weekly salary between two different between pay rates, but you will only have a few minutes to make a decision. These are the choices:

Plan I: Initial salary of $10,000 with an increase of $500 per week
Plan II: Initial salary of $10,000 with an increase of 5% per week

a. Without applying any math formulas, which plan would you select to get a higher pay rate? Explain your decision.

b. Make a table of values for your weekly salary under Plan I.

Week	Plan I
0	10,000
1	10,500
2	
5	
10	
15	
20	

c. Make a table of values for your weekly salary under Plan II.

Week	Plan II
0	10,000
1	10,500
2	
5	
10	
15	
20	

d. Was your initial decision the correct one? Why or why not?

e. Without graphing, determine which of the two plans illustrates a linear growth and which illustrates an exponential growth. Justify your decision.

f. Using w for weeks, find a function, $L(w)$, for the plan that reflects a linear growth.

g. Using w for weeks, find a function, $E(w)$, for the plan that reflects an exponential growth.

h. Graph both functions using the window [0, 80, 10] by [0, 100000, 10000]. Show a rough sketch of your graphs.

i. Comment on both growth patterns.

8. The average price of gas in March 2005 in Kuwait was \$0.78 per gallon. In March 2008, it was \$0.90 per gallon. Use the equation $P(t) = P_0 e^{kt}$ to find the following.
(Hint: Let $t = 0$ correspond to 2005.)
Source: http://money.cnn.com

a. Write a growth function for the gas prices in Kuwait.

b. Find the value of k to 3 decimal places. Show all your work.

c. Write the equation $P(t) = P_0 e^{kt}$ using your answers from parts (a) and (b).

d. Complete the table below to predict the price of gas in Kuwait in 2012. Round the prices to 2 decimal places.

t	3	4	5	6	7	8
$P(t)$						

e. Graph the function using the window [-10, 15, 2] by [0.48, 1.6, 0.1] and use your graph to estimate the year when the gas price will reach $1.47 per gallon.

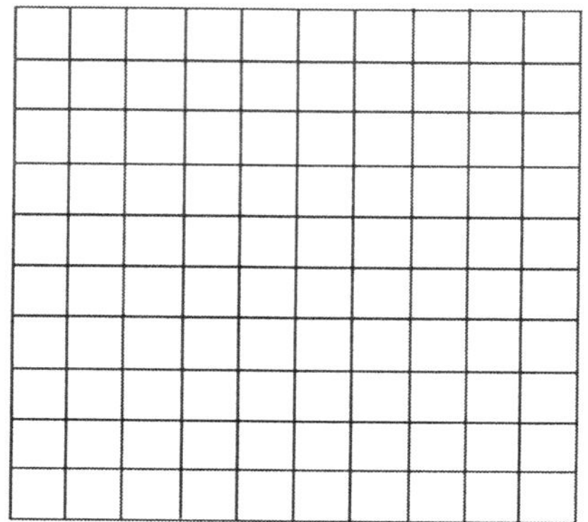

VIII. Rational Functions and Applications

Graphs of Rational Functions

1. Write a rational function that has the following conditions.
Note: Your answers may vary from those of your classmates.

a. The rational function has a horizontal asymptote at $y = 0$.

b. The rational function has no horizontal asymptote.

c. The rational function has a horizontal asymptote at $y = \frac{3}{2}$.

2. Using a complete sentence, explain how to find the vertical asymptote(s) of a rational function.

3. Given the rational function $f(x) = \frac{6x+14}{2x+4}$, answer the following:

a. Find the horizontal asymptote, if one exists.

b. Find any vertical asymptote(s).

c. Graph the function; denote any asymptotes with dashed lines.

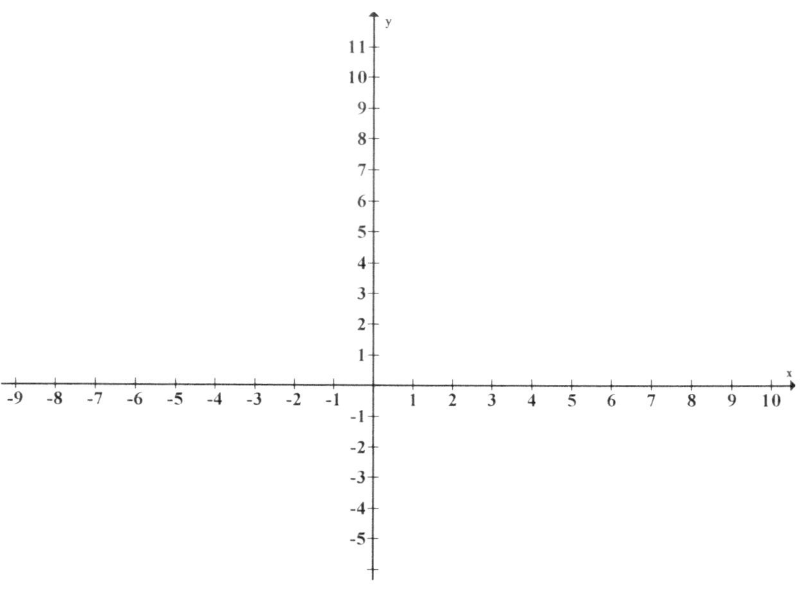

4. LCD (liquid-crystal display) TVs are becoming popular across the nation. Suppose the cost of a new 42-inch LCD TV to replace an old TV is $2,000. The new 42-inch LCD would add an average of $55 to the annual electric bill.
Source: http://consumerreports.org

a. Considering only the cost and annual electric bill and assuming that the 42-LCD TV has a lifespan of 7 years, the total annual cost, $C(x)$, for the new TV for x number of years could be modeled by the rational function $C(x) = \dfrac{2000 + 55x}{x}$. Use the function to determine the total annual cost of a 42-LCD TV that lasts 7 years. Answer to the nearest dollar.

b. Using your calculator, select an appropriate viewing window and graph the function. Label the axes.

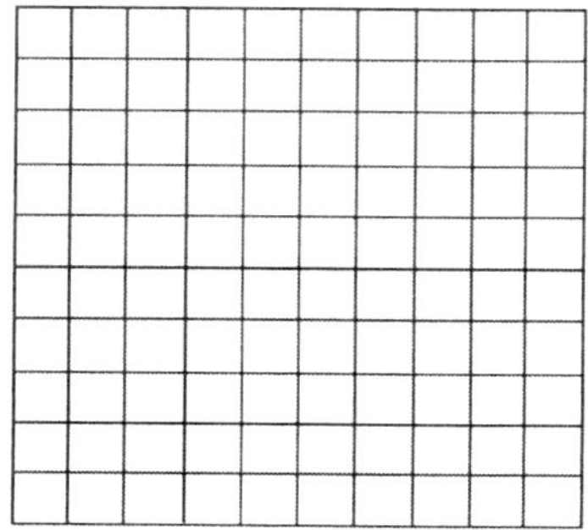

c. What does the graph show about the cost?

d. Find the equation of the vertical asymptote of the function.

e. Find the equation of the horizontal asymptote of the function.

f. State the significance of the horizontal asymptote in the context of the problem.

g. If the total annual cost is $255, how many years has the LCD TV lasted? Solve algebraically and corroborate your answer with the graph.

5. On June 2008, the United States Bowling Congress announced that it will hold 13 Championships from 2011 through 2029 in a privately funded bowling center on land being made available at Disney's Wide World of Sports complex in Osceola County, Florida. Each tournament will run for approximately 20 weeks and is expected to draw tens of thousands of bowlers and spectators (*source: http://bowl.com*). Patty and Linda own the Strike & Spares Designs bowling apparel Web store. They plan to design and market bowling shirts to promote and commemorate the inaugural tournaments. The total costs to begin this project are $600 for material plus $5.75 to design and print each shirt.

a. Write a linear function for the total cost, C, of producing x number of shirts.

b. Write a rational function for the *average* cost, A, of producing x number of shirts. (Hint: Use your linear function to help in your calculations.)

c. Graph the average function in the window [0, 500, 50] by [0, 80, 10]. Show a rough sketch of your graph.

d. Use your graph to find the average cost of producing 100 shirts.

e. If the average cost is $8.15, how many bowling shirts can they produce? Solve algebraically or graphically.

f. What is the horizontal asymptote of the average cost function? State its meaning in the context of this problem.

Domain and Range

1. A collaborative learning group in a college algebra class is reviewing rational functions. Dario's group is working on comparing the domains for the rational functions $f(x) = \frac{6+7x}{x}$ and $f(x) = 7 + \frac{6}{x}$. Dario says the domains are exactly the same, but one of his classmates alleges that the domains are different.

a. Who is correct and why? State the domain.

b. Graph both functions with your calculator and corroborate your answer from part (a). Include a rough sketch of your graphs; state the viewing window.

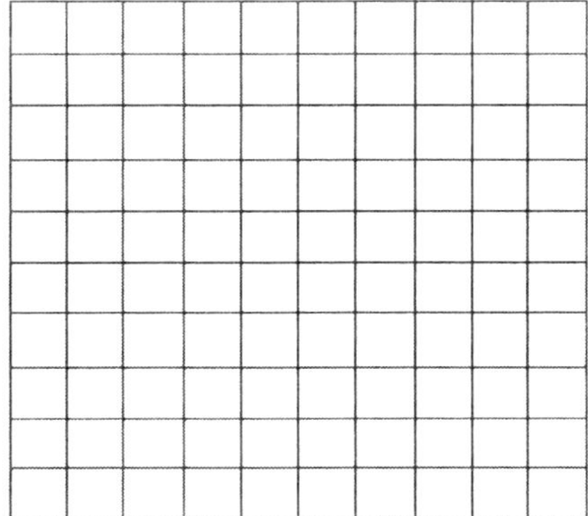

c. Find the vertical and horizontal asymptotes.

2. The average cost, A, of producing x number of shirts at the Strike & Spares Designs bowling apparel Web store is modeled by:

$$A(x) = \frac{600 + 5.75x}{x}$$

Find the domain of this rational function and state its meaning in the context of this problem.

3. Use the following graph to answer the questions below:

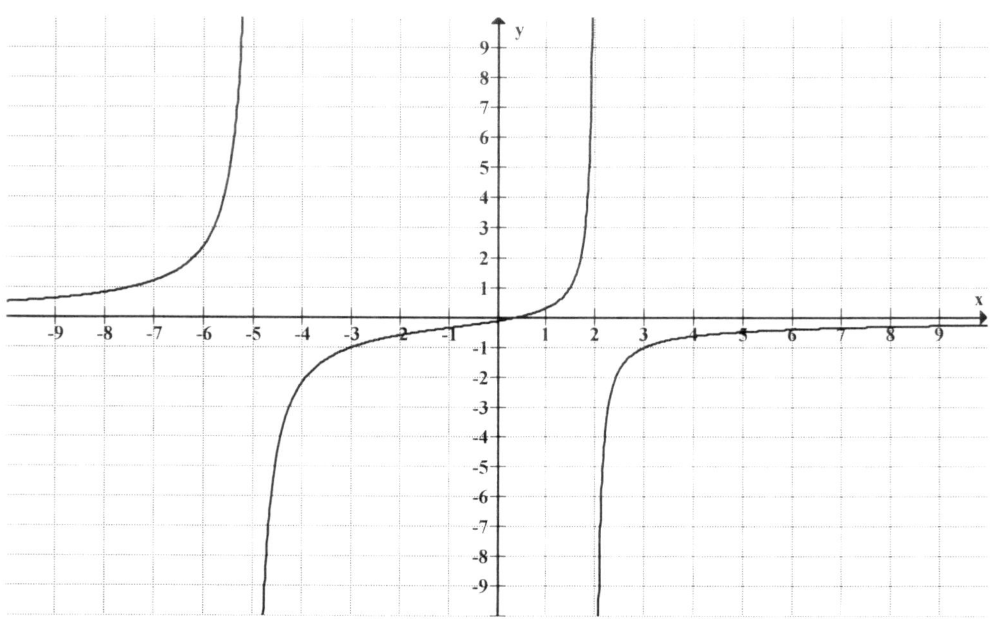

a. Find any vertical asymptote(s).

b. Find any horizontal asymptote(s).

c. State the domain in interval notation.

d. State the range in interval notation.

e. Is it possible for the graph of a rational function to cross a horizontal asymptote? A vertical asymptote? Explain your answers.

4. Given the function $f(x) = \dfrac{6x - 2}{-2x^2 - 6x + 20}$, answer the following:

a. Find the domain *algebraically*. Show all work.

b. Graph the function and compare it with the graph from problem (3). Comment on the relationship between the vertical asymptotes and the domain.

Appendix: Graphing Calculator Help for TI-83, TI-83 Plus, TI-84, TI-84 Plus

Appendix A

Graphing Lines
(Models TI-83, TI-83 Plus, TI-84, TI-84 Plus)

Example 1: Graph the equation $y = 2x + 1$

We start with the [Y=] key, located in the upper-left corner of the calculator. Enter $2x + 1$ into Y1. If there are any previous equations on the calculator, erase them by pressing [CLEAR]. Note: Use the [X,T,Θ,n] key to type the x variable.

This is the corresponding screen:

```
Plot1 Plot2 Plot3
\Y1=2X+1
\Y2=
\Y3=
\Y4=
\Y5=
\Y6=
\Y7=
```

[Note: If you see any "Plot" highlighted, move the cursor up to it and press [ENTER] to deactivate it; no Plot should be activated.]

Now that we have the equation in the calculator, we are ready to graph it. First, we must make sure that we use an appropriate viewing rectangle (window). The standard window will have values of X ranging from -10 to 10, and the same values for Y. That is:

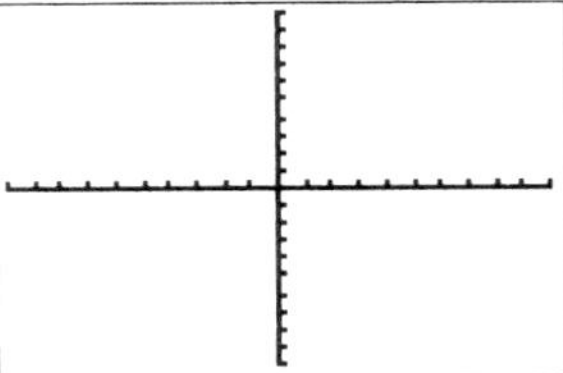

Press the [WINDOW] key. Xmin is the minimum x-value along the x-axis, and Xmax is the maximum x-value; Ymin is the minimum y-value along the y-axis, and Ymax is the maximum y-value.

Enter the following values

Xmin = -10 (remember to use the small negative sign, [(-)], at the bottom of your calculator, ***not*** the minus sign [-])
Xmax = 10
Xscl = 1 (Xscl sets the distance between tick marks on the x-axis)
Ymin = -10
Ymax = 10
Yscl = 1 (Yscl sets the distance between tick marks on the y-axis)
Xres = 1 (This sets the pixel resolution; choose 1 for a good graph)

This is the corresponding screen:

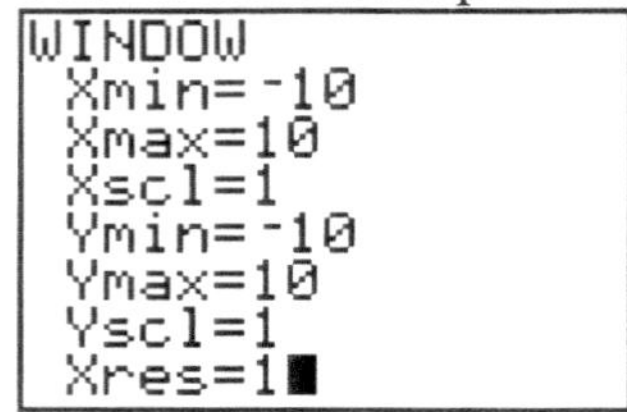

Information about the viewing window can be written in a shorten version, like this: [Xmin, Xmax, Xscl] by [Ymin, Ymax, Yscl]. In our example: [-10, 10, 1] by [-10, 10, 1].

Now, press GRAPH to see the graph of Y1 = $2x + 1$. It should look like this:

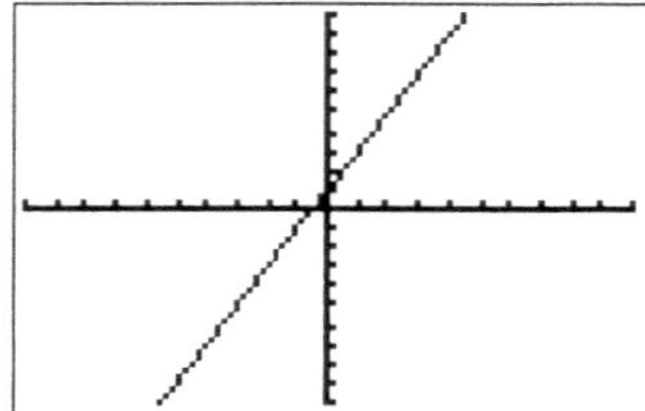

Notice how the graph shows both x- and y-intercepts—that is, the points where the line crosses the x and y axes. In this case, the line crosses the y-axis at (0,1).
The x-intercept is (- ½, 0).

We can verify this algebraically:

If $x = 0$ → $y = 2(0) + 1$

$y = 1$

So, the y-intercept is, indeed, (0,1).

If $y = 0$ → $0 = 2x + 1$

$x = -\frac{1}{2}$

So, the x-intercept is, indeed, (- ½, 0).

Example 2: Graph $y - 12 = 4x$ with the graphing calculator.
In this case, we must first isolate the y before we enter the equation into the graphing calculator.
This results in: $y = 4x + 12$

Just as we did before, enter $4x + 12$ into Y1 on your calculator.

(Remember to erase any previous equation on the calculator, by pressing CLEAR.)

```
Plot1 Plot2 Plot3
\Y1■4X+12
\Y2=
\Y3=
\Y4=
\Y5=
\Y6=
\Y7=
```

To access the standard window [-10, 10, 1] by [-10, 10, 1] using a shortcut, we can simply press [ZOOM] and select option 6, "standard." This will generate the graph automatically.

When we graph this equation by using the standard window, it will look like this:

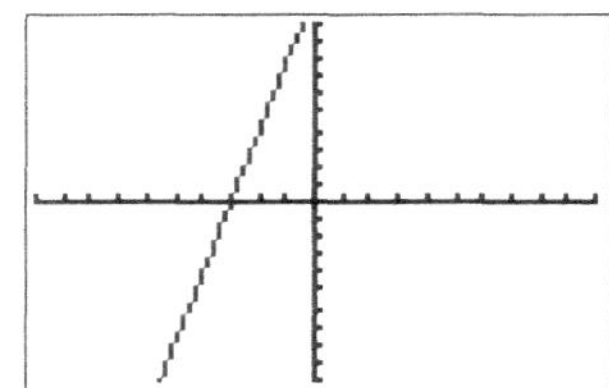

Notice that we cannot see the y-intercept. Since this line crosses the y-axis at the point (0,12), which is beyond our standard viewing window of [-10, 10, 1] by [-10, 10, 1], we must assign a larger value to Ymax.

Press the [WINDOW] key and change Ymax to 15, or any larger value, until you clearly see the y-intercept.

The new graph will look similar to this:

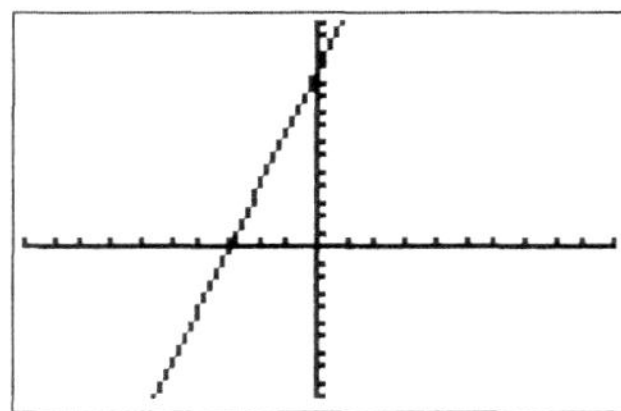

Now the graph shows both x- and y-intercepts, so this graph is OK.

If we find the intercepts of $y = 4x + 12$ algebraically, the coordinates will be:
y-intercept: (0,12) x-intercept: (-3, 0)
The graph shows both of these values.

Now you try it!
Graph the following lines with your graphing calculator. For each graph, find both x- and y-intercepts algebraically and make sure that the graph shows these values.

1. $y = 3x + 8$
2. $3y - 6x - 2 = 16$ (Remember to isolate the y!)
3. $y = -4x - 28$ (You may have to adjust your window here.)

Appendix B

Linear Regression
(Models TI-83, TI-83 Plus, TI-84, TI-84 Plus)

Not all data lie on a straight line. In the real world, changes over a given period of time are not always constant. For example, a stock may rise steadily over a year's time and then increase or decrease rapidly, as in Apple's expansive growth after introducing the iPod.

If we were to plot certain data over a given period of time, we would see that the data would not fit on a straight line because of fluctuations with increases and decreases. We could, however, make future predictions by finding a linear equation that "best fits" the data. This equation is not exact, but it will allow us to obtain a pretty good approximation as long as the changes in data do not vary too greatly. Linear approximations, or *linear regressions*, allow us to make predictions that are easy and fairly accurate as long as the data behave rather linearly. Simple linear regression allows us to generate an equation for the line that most nearly fits the given data.

Making Scatterplots

- Set your window according to the given data: Set the Xmin and Xmax according to the smallest and largest x-values given. Do the same for the Ymin and Ymax.
- Go to [STAT], choose option 1, Edit, and enter the data under L1 and L2. If data are already entered, place your cursor over the L1 or L2 using the up arrow key, hit [CLEAR] and then [ENTER]. This will delete all previous data in that list.
- Go to [Y=] to delete or turn off all previous equations.
- Turn on the Plot1 under the [Y=] key, then press graph.
- From here you can use the trace key and your left and right arrows to see the data values from the scatterplot.

Make sure you turn the Plot off when you are done!

Example:

Enter the data from the table below into lists and graph the scatterplot of the data using the window [0, 8] by [0, 312].

x	1	2	3	4	5
y	47	72	74	182	309

Following are the corresponding screens:

```
WINDOW
 Xmin=0
 Xmax=8
 Xscl=1
 Ymin=0
 Ymax=312
 Yscl=1
 Xres=1
```

```
Plot1 Plot2 Plot3
\Y1=
\Y2=
\Y3=
\Y4=
\Y5=
\Y6=
\Y7=
```

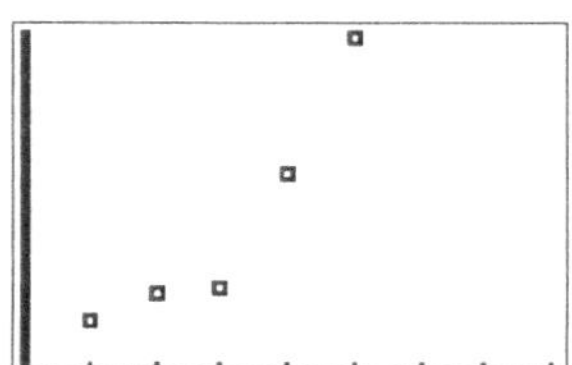

To find a linear function that best fits a given data set (linear regression):

- Enter the data under L1 and L2.
- Choose [STAT], scroll right to [CALC] and choose option 4 "LinReg(ax+b)," then press [ENTER].

The linear equation $y = ax + b$, where the values of a and b are given, denotes the linear regression line.

- The r value—that is, the *correlation coefficient*—measures the goodness of the fit of the line with the data points.
- If you do not see an r value, the Diagnostic feature is not turned on. If this is the case, go to "catalog" (the second function above the number 0) and scroll down until you see "DiagnosticOn."
- Press [ENTER].
- The closer the $|r|$ is to 1, the closer the data points fit on the regression line.
- If $r = 0$, there is no linear relationship.
- If r is positive, the slope of the line is positive.
- If r is negative, the slope of the line is negative.

Now you try it!

The following table shows Apple's net profit (in billions of dollars) for its first quarter results for the years 2004 to 2008. Let x be the number of years after 2004.

Year	2004	2005	2006	2007	2008
Net Profit	.063	.295	.565	1.00	1.58

Source: http://www.apple.com

a. Construct a scatterplot using the data from the table.

b. Find an equation for your regression line.

c. State the slope and interpret it.

d. What does the model estimate for the first quarter net profit in 2009?

e. Observing the data in the table above, estimate the year in which Apple launched its popular iPod product.

Appendix C

Graphical Solutions (Finding Intersections)
(Models TI-83, TI-83 Plus, TI-84, TI-84 Plus)

How do we solve equations graphically?

Graphical solutions are usually found by setting the left side of the equation equal to Y1 (in a graphing calculator) and setting the right side of the equation equal to Y2. Then, we find the point of intersection of both graphs.

Example: Solve the equation $2x + 1 = 7$ graphically.

We start with the [Y=] key. Enter the left side, $2x + 1$, into Y1 on your graphing calculator (remember to [CLEAR] any previous equations in the calculator); enter the right side, 7, into Y2.

This is the corresponding screen:

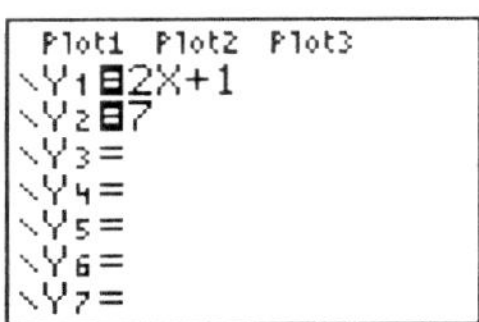

Now we graph both equations, making sure that we use an appropriate viewing rectangle (window).

The standard window has values of X ranging from -10 to 10, and the same values for Y. Let's review how to set the window:

Press [WINDOW]. Xmin is the minimum x-value along the x-axis, and Xmax is the maximum x-value; Ymin is the minimum y-value along the y-axis, and Ymax is the maximum y-value.

Enter the following values
Xmin = -10
Xmax = 10
Xscl = 1
Ymin = -10
Ymax = 10
Yscl = 1
Xres = 1

To access the standard window [-10, 10, 1] by [-10, 10, 1] we can also simply press [ZOOM] and select option 6, "standard." This will generate the graph automatically.

This is the corresponding screen:

```
WINDOW
 Xmin=-10
 Xmax=10
 Xscl=1
 Ymin=-10
 Ymax=10
 Yscl=1
 Xres=1
```

Now, press GRAPH to see the graphs of Y1 = $2x + 1$, and Y2 = 7. Your graphs should look like this:

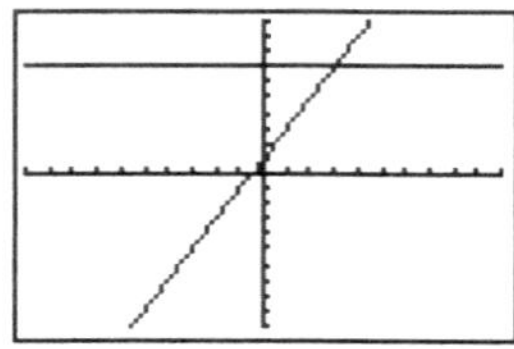

The solution to our equation will be the x-coordinate of the point of intersection. To find the point of intersection for the two lines, Y1 = $2x + 1$ and Y2 = 7, we press the keys 2nd TRACE to access the CALCULATE menu. This is the resulting screen:

```
CALCULATE
1:value
2:zero
3:minimum
4:maximum
5:intersect
6:dy/dx
7:∫f(x)dx
```

Choose option 5, "intersect." When the calculator prompts for the first curve, simply press ENTER three times.

The coordinates of the point of intersection will be shown at the bottom of the screen:

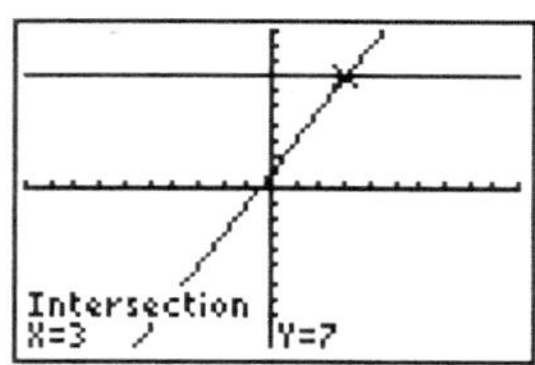

Since we are seeking for an x-value that satisfies $2x + 1 = 7$, we only need to select the x-coordinate, 3. We ignore the y-value because our equation contains only the variable x. The solution to $2x + 1 = 7$ is $x = 3$.

Solving this equation algebraically (symbolically), we would get the same answer:

$2x + 1 = 7$

$2x + 1 - 1 = 7 - 1$

$x = 3$

Verifying: $2(3) + 1 = 7$ It checks!

Another example:
Solve the equation $9 - x = 3x + 1$ graphically.

Just as we did before, enter $9 - x$ into Y1 on your calculator, then enter $3x + 1$ into Y2:

Y1 = $9 - x$
Y2 = $3x + 1$

The standard window is usually a good starting point for graphing, although sometimes we may have to adjust the viewing window for larger or smaller values.

Your graphs should look like this:

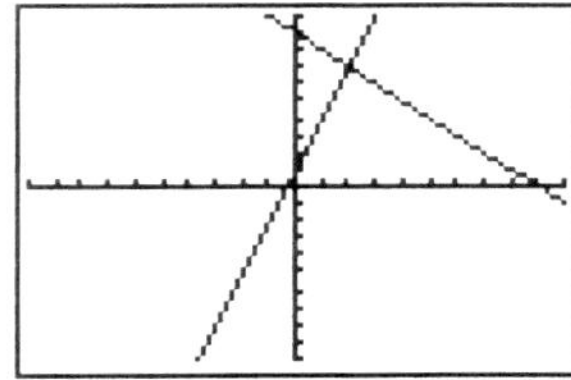

Now, we find the point of intersection for the two lines, Y1 = $9 - x$ and Y2 = $3x + 1$, by pressing the following keys:

[2nd] [TRACE] and selecting option 5, "intersect."

Remember: When the calculator prompts for the first curve, simply press [ENTER] three times.

The coordinates of the point of intersection will be shown at the bottom of the screen:

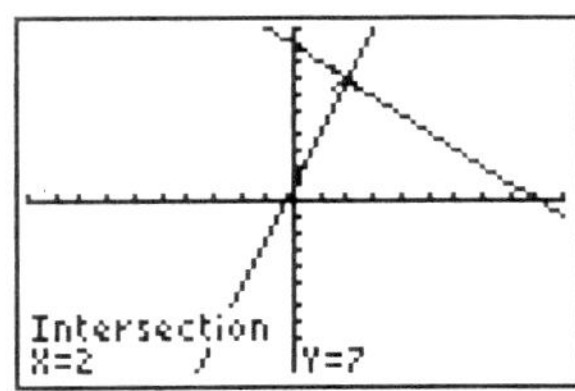

Since we are seeking for an x-value that satisfies $9 - x = 3x + 1$, we only need to select the x-coordinate, 2. Therefore, the solution to $9 - x = 3x + 1$ is $x = 2$.

Solving this equation algebraically, we will get the same answer, $x = 2$ (try it!).

There are advantages and disadvantages to graphical solutions. One advantage is that we can solve equations using technology, and we can gain a better understanding of what it means to solve an equation by looking at the point of intersection. One disadvantage of solving equations with graphs is that sometimes it is hard to get exact values. The solution is often an estimate; this is especially common when we deal with decimal/fractional answers.

Let's practice more examples:

Solve $\frac{1}{2}x = (\frac{1}{3})(x - 4)$ graphically by studying the given graphs:

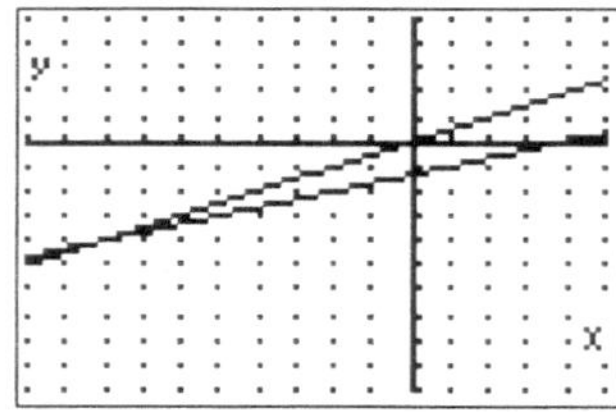

X min = -10, X max = 5, X scale = 1
Y min = -10, Y max = 5, Y scale = 1

This time both graphs are given. There is no need to graph the equations. The coordinates of the point of intersection are (8, - 4). Therefore, the answer is x = -8.

We can verify: $(\frac{1}{2})(-8) = (\frac{1}{3})(-8 - 4)$

$-4 = (1/3)(-12)$

$-4 = -4$

Solve graphically: $3(4 - 2x) = 4x + 12$

Solution: Let Y1 $= 3(4 - 2x)$
Y2 $= 4x + 12$

When we graph these equations by using the standard window, we get:

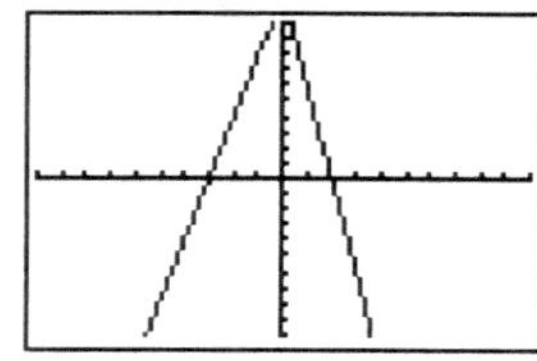

Notice that we cannot see the intersection of these lines. Since the second line (Y2 $= 4x + 12$) crosses the y-axis at the point (0,12), which is beyond our standard viewing window of [-10, 10, 1] by [-10, 10, 1], we must assign a larger value to Ymax.

Press the WINDOW key and change Ymax to 15, or any larger value, until you clearly see the intersection.

The new graph will look similar to the following:

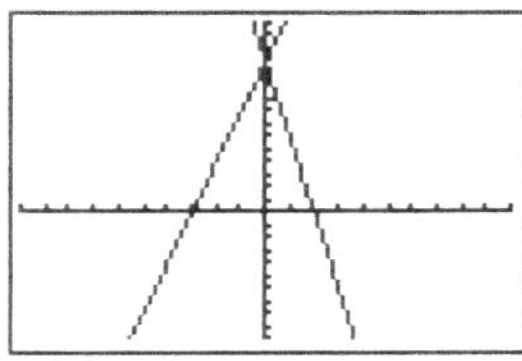

The coordinates of the point of intersection, shown at the bottom of the screen are:

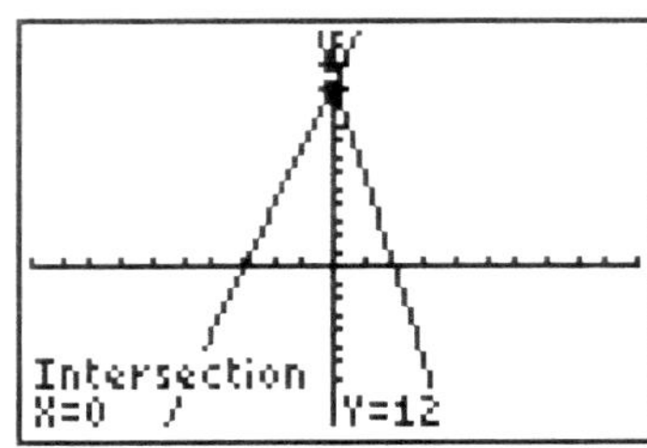

Therefore, the solution to $3(4 - 2x) = 4x + 12$ is $x = 0$.

Of course, solving algebraically we would get the same answer:

$$3(4 - 2x) = 4x + 12$$
$$12 - 6x = 4x + 12$$
$$-6x - 4x = 12 - 12$$
$$-10x = 0$$
$$x = 0$$

Now you try it!

Solve graphically and algebraically:

a. $3x - 5 = x + 1$

b. $2x - 2 = -4x - 14$

c. $9(x - 1) = 27$

Appendix D

Creating a Table of Values
(Models TI-83, TI-83 Plus, TI-84, TI-84 Plus)

Given the equation $y = 3x - 72$, we know that y is the dependent variable (its value depends on the value of x) and x is the independent variable. To create a table of values that satisfy this equation, we can just pick a value for x, substitute it into the equation, and solve for y.

Example: If $x = -2$,
$y = 3(-2) - 72$
$y = -6 - 72$
$y = -78$

If $x = -1$,
$y = 3(-1) - 72$
$y = -3 - 72$
$y = -75$ and so on.

A sample table for this equation is:

x	y
-2	-78
-1	-75
0	-72
1	-69
2	-66

Some tables do not require the use of a calculator, since we work with small numbers. Nevertheless, knowing how to work with a graphing calculator will make it easier to make tables for more complicated equations. How can we create a table of values with the graphing calculator?

Let us generate the previous table with the graphing calculator. We will make a table of ordered pairs that satisfy the equation $y = 3x - 72$, for integer values of x beginning at -2.

1. We first press the [Y=] key, and enter the equation $y = 3x - 72$ under Y1. You can find the [Y=] key on the upper left corner of the calculator. (Press [CLEAR] to erase any equations that were previously used.) Your screen will look like this:

```
Plot1 Plot2 Plot3
\Y1=3X-72■
\Y2=
\Y3=
\Y4=
\Y5=
\Y6=
\Y7=
```

2. Press [2nd] [TBLSET]
(This is the second function associated with the [WINDOW] key.)

You will see a similar screen:

3. Under TblStart enter the beginning value: (-)2
   (Note: Use the negative sign, [(-)], at the bottom of your calculator, *not* the minus sign [-].)

4. The symbol △Tbl = 1 means adding 1 to the preceding x value
   (Note: if you enter △Tbl = -1 , the values will decrease by 1.)

5. | | | | | |
   |---|---|---|---|---|
   | Indpnt: | Auto | Ask | (these are the x values) | → choose Auto |
   | Depend: | Auto | Ask | (these are the y values) | → choose Auto |

   This is the corresponding screen:

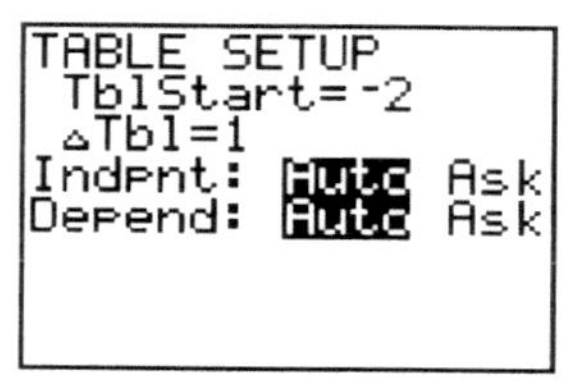

6. Press [2nd] [TABLE] to see your new table.
   (This is the second function associated with the [GRAPH] key.) Notice these are the same values we found before.

   You can use the arrows to scroll up and down the table.

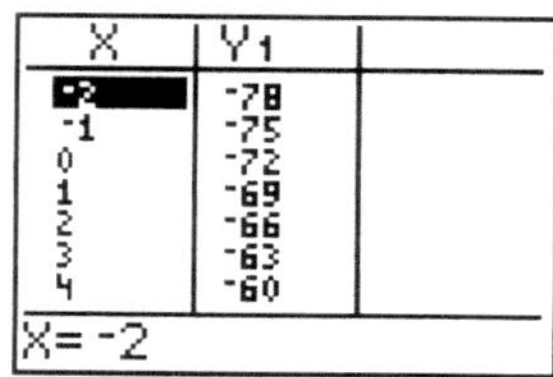

<u>Now you try it!</u>
Make a table of ordered pairs that satisfy the equation $y = 4x - 60$, for integer values of x beginning at -5.

Note: You can choose to supply the x-values yourself by setting Indpnt to *Ask*. Then, you can press [2nd] [TABLE]. The calculator will display $x =$ and you can type your values and press [ENTER] after each one.

Appendix E

Numerical Solutions
(Models TI-83, TI-83 Plus, TI-84, TI-84 Plus)

A table of values is also called a *numerical representation* of the corresponding equation. The table feature in the graphing calculator displays a partial numerical representation, where we can scroll up or down to see more values.

How do we solve equations numerically?

First example: Solve the equation $2x + 1 = 11$ numerically.

We begin by creating a table for the expression $2x + 1$.
(See Appendix D, Creating a Table of Values)

Enter $2x + 1$ into Y1 on your calculator and make a table of values, starting at an estimate value. Since the expression $2x + 1$ is equal to 11, a small number, we can start the table at 0 or any smaller number, and choose increments of 1.
The resulting table will be:

X	Y1
0	1
1	3
2	5
3	7
4	9
5	11
6	13
X=0	

Notice that Y1 (which we used for $2x + 1$) equals 11 when $x = 5$. Therefore, the solution to the equation $2x + 1 = 11$ is $x = 5$.

Solving this equation algebraically (also called *symbolically*), we would get the same answer:

$2x + 1 = 11$
$2x + 1 - 1 = 11 - 1$
$2x = 10$
$x = 5$

Verifying: $2(5) + 1 = 11$. It checks.

Another example:
Solve the equation $9 - x = 3x + 1$ numerically (with a table).

Both sides of the equation contain the variable x.

To solve this equation numerically, we begin by creating a table for the expressions: $9 - x$ and $3x + 1$.

Just as we did before, enter $9 - x$ into Y1 on your calculator, then enter $3x + 1$ into Y2, and make a table of values starting at an estimate value. Since the expressions are not too large, we can start the table at, let's say 0, and choose increments of 1.

The resulting table will be:

X	Y1	Y2
0	9	1
1	8	4
2	7	7
3	6	10
4	5	13
5	4	16
6	3	19

X=0

Notice that Y1 = Y2 when $x = 2$. Since we used Y1 for $9 - x$ and Y2 for $3x + 1$, then $9 - x = 3x + 1$ when $x = 2$. Thus, $x = 2$ is the solution to this equation.

Verifying: $9 - (2) = 3(2) + 1$

$7 = 7$

Solving this equation algebraically (symbolically), we get the same answer.

There are advantages and disadvantages to numerical (table) solutions. One advantage is that we can solve equations using technology, which is a great help for computations. Also, we can gain a better understanding of what it means to solve an equation by looking at the numerical solution. One disadvantage of solving equations with tables is that sometimes it is hard to get exact values, and we can only estimate the numerical solution; this is especially common when we deal with decimal/fractional answers.

Let's practice more examples:

Solve $\frac{1}{2}x = (\frac{1}{3})(x - 4)$ numerically by studying the corresponding table of values:

X	Y1	Y2
-10.00	-5.00	-4.67
-9.00	-4.50	-4.33
-8.00	-4.00	-4.00
-7.00	-3.50	-3.67
-6.00	-3.00	-3.33
-5.00	-2.50	-3.00
-4.00	-2.00	-2.67

X= -10

This time the table is given, so there is no need to do it yourself. Looking at this table, we know that Y1 corresponds to the values for the left side of our equation, $\frac{1}{2}x$, and Y2 corresponds to the values for $(\frac{1}{3})(x - 4)$. Since Y1 is equal to Y2 when $x = -8$, the solution to the given equation is $x = -8$.

We can verify: $(\frac{1}{2})(-8) = (\frac{1}{3})(-8 - 4)$

$$-4 = (1/3)(-12)$$
$$-4 = -4$$

Solve the following equation numerically and algebraically: $3(4 - 2x) = 4 + 2x$

Solution:

Let $Y1 = 3(4 - 2x)$

$Y2 = 4 + 2x$

Starting the table at $x = -1$, we get:

X	Y1	Y2
-1	18	2
0	12	4
1	6	6
2	0	8
3	-6	10
4	-12	12
5	-18	14

X= -1

Where is Y1 = Y2? When $x = 1$, both Y1 and Y2 equal 6. Therefore, the numerical solution to the equation $3(4 - 2x) = 4 + 2x$ is $x = 1$

Solving algebraically:

$$3(4 - 2x) = 4 + 2x$$
$$12 - 6x = 4 + 2x$$
$$-6x - 2x = 4 - 12$$
$$-8x = -8$$
$$x = 1$$

<u>Now you try it!</u>

Solve numerically and algebraically:

a. $3x - 5 = x + 1$

b. $2x - 2 = -4x - 14$

c. $9(x - 1) = 27$

Appendix F

Finding Maximum or Minimum Points (Vertex)
(Models TI-83, TI-83 Plus, TI-84, TI-84 Plus)

To find maximum and minimum points on a graph, use the CALCULATE feature by pressing [2nd] [TRACE] and choosing option 3 for minimum or option 4 for maximum.

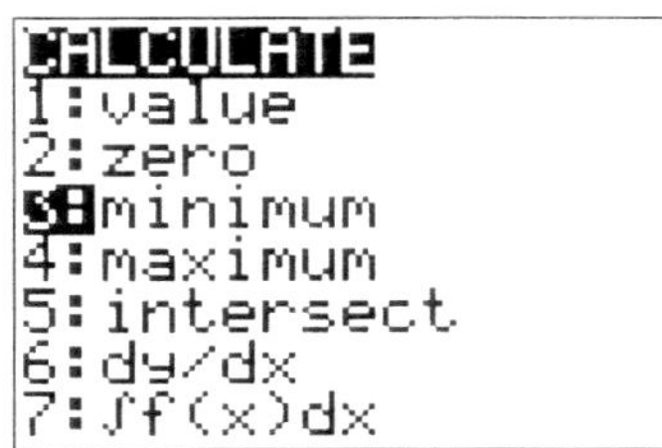

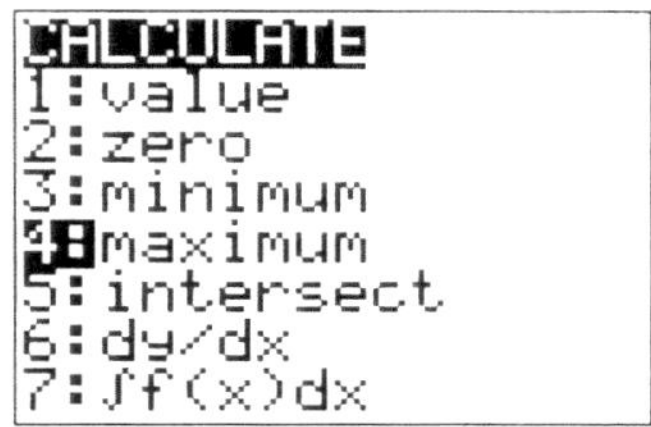

Example:
Graph the parabola $y = x^2 - 4x + 7$ using the standard window. Find the minimum point of this graph (we know we are looking for a minimum because this parabola opens upward). Once you graph this parabola, press [2nd] [TRACE] and select option 3.

First, we need to specify an interval containing the desired low point on the graph by setting the bounds of the function. Since we wish to find the minimum, when the question "Left Bound?" appears at the bottom of the screen, use the right arrow, [▸], repeatedly to move the blinking cursor to the left of the relative minimum. Press [ENTER].

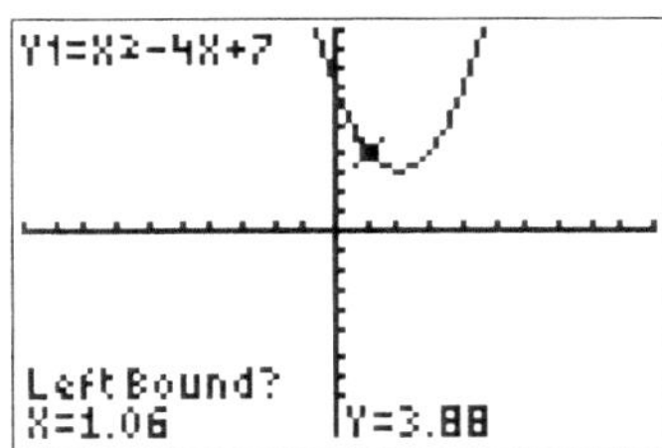

When the question "Right Bound?" appears at the bottom of the screen, press [▸] repeatedly to move the blinking cursor to the right of the relative minimum.
Press [ENTER].

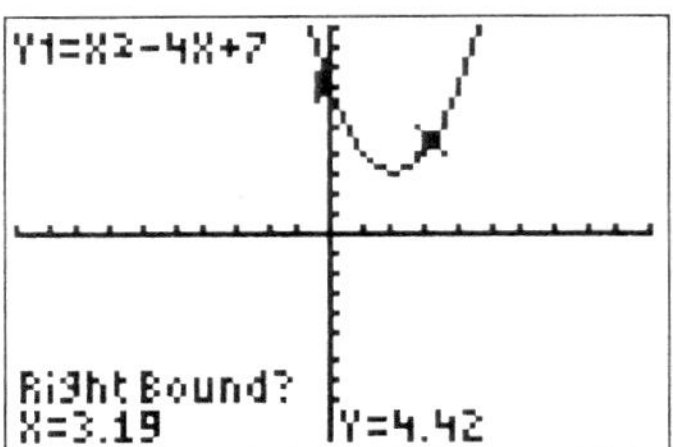

The arrows at the top of the screen indicate the boundaries between which the calculator will give the relative minimum. (The arrows must point toward each other.) The question "Guess?" appears at the bottom of the screen; locate the cursor between the established boundaries and press [ENTER] to display the minimum value.

See below for the corresponding screens:

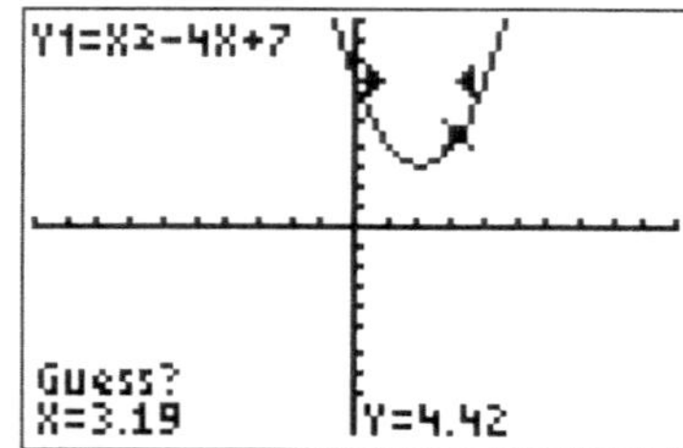

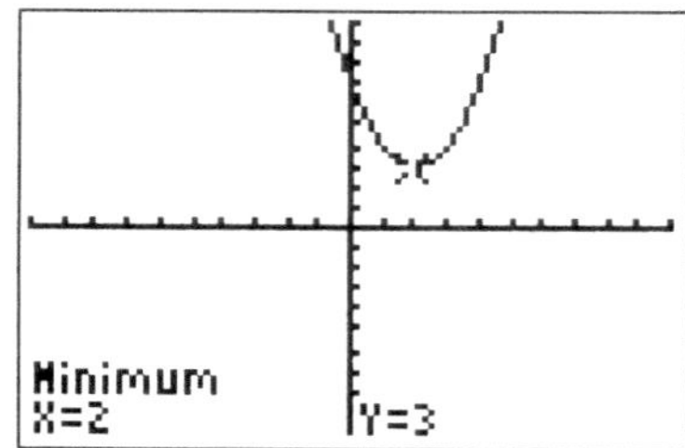

Now you try it!

a. Find the vertex (maximum or minimum) for $f(x) = -x^2 + 3x - 4$ with the graphing calculator.

b. Find the vertex algebraically for $y = -5x^2 + 20x + 3$ using the $x = \dfrac{-b}{2a}$ formula.

Then, find the vertex with your grapher and compare your answers.

Appendix G

Finding Zeros
(Models TI-83, TI-83 Plus, TI-84, TI-84 Plus)

The *zero* function on your grapher can be used to find the x-value where the graph of a function crosses the x-axis, or the x-intercept. This is a useful tool in solving equations. We can solve an equation by moving all the terms to one side of the equation, leaving 0 on the other. Then we enter the nonzero side of the equation into the calculator using the [Y=] key.

Example:
Find the solutions (zeros) for $(x - 3.6)(x + 5.9) = 0$

We know from the zero product property what our answers are, but let's find them with the calculator to give us a visual aspect of the solutions of this equation.

First, we graph the equation:

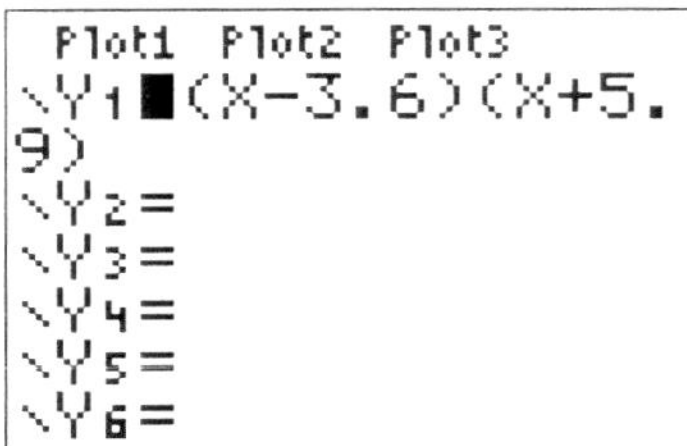

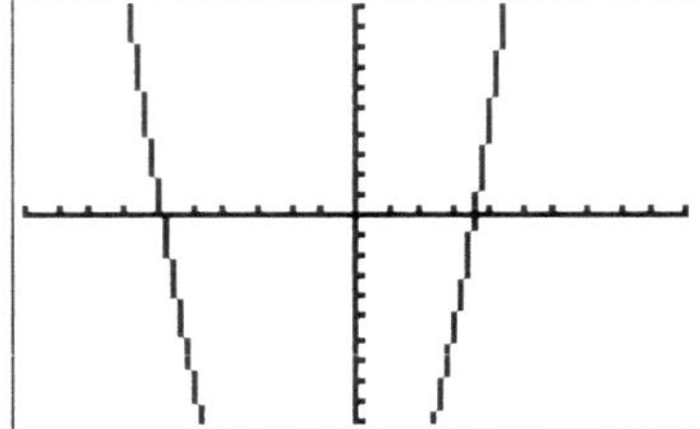

Normally, we would adjust our viewing window to display the vertex of the parabola, but since we are just looking for the solutions of the equation $(x - 3.6)(x + 5.9) = 0$, which are the x-intercepts (zeros), this window would be acceptable.

We find the x-intercept(s) of the graph by establishing a left bound and a right bound around the point on the x-axis (using the left and right arrows).
Press [2nd] [TRACE] to access the CALCULATE feature and select option 2 (zero).

CALCULATE
1:value
2:zero
3:minimum
4:maximum
5:intersect
6:dy/dx
7:∫f(x)dx

Since a graph may have more than one x-intercept, specify an interval containing the desired x-intercept. Let's say that we want to find the x-intercept to the left.
When the question "Left Bound?" appears at the bottom of the screen, keep moving the cursor with the left arrow, [◄], until it is on the left of the zero or x-intercept. When you are asked for the right bound, keep moving the [►] arrow key until your cursor is to the right of the zero.

See the screens below:

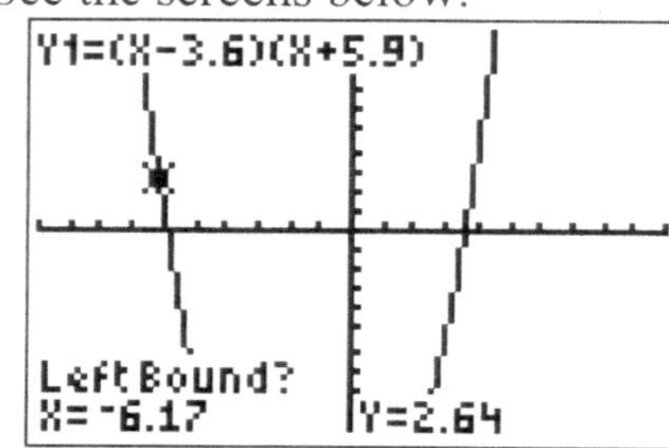

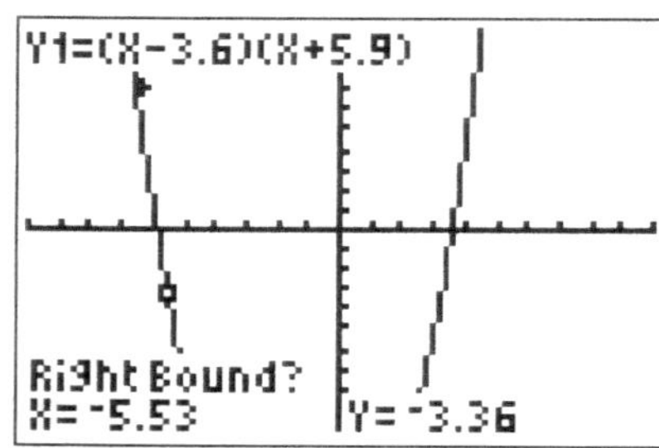

When the calculator asks "Guess?" just press [ENTER] to get the x-intercept or zero.

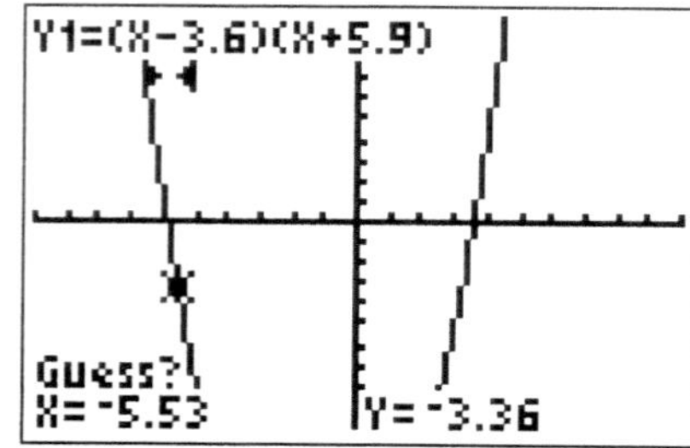

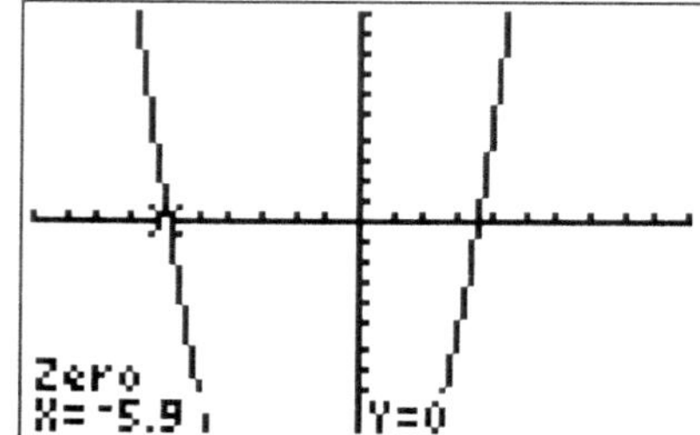

This value will be one of the solutions of the given equation.
Now try finding the other zero on your own with the calculator.
You should get $x = 3.6$.

Now you try it!
Find the zeros and the vertex for the following, and round your answers to 3 decimal places where possible.

a. $y = x^2 - 6.8x + 6.56$

Zeros ____________ vertex ______________

b. $f(x) = -x^2 + 7.56x - 6.6$

Zeros ____________ vertex ______________

c. $y = 2x^2 - 7.3x - 2.9$

Zeros ____________ vertex ______________

Activity Workbook with
Real-World Applications for College Algebra

Evaluation Form

Please take a few minutes to fill out this evaluation. Your comments will help us maintain the quality and applicability of this workbook. Be as concise and specific as possible.

1. Which activities did you find most engaging?

2. Which activities did you find least engaging?

3. Are there any activities/worksheets/problems that need further explanation? If so, please elaborate.

4. Please suggest other topics to include in this workbook.

5. What suggestions would you offer to improve the overall quality of this workbook?

6. Would you recommend this workbook to others? Why or why not?

Please send your evaluation to:
Gisela Acosta-Carrasquillo or Margie Fernandez-Karwowski
Valencia Community College
East Campus, MC 3-16
701 N Econlockhatchee Trail
Orlando, FL 32825

You may also e-mail your evaluation/comments to:
gacosta@valenciacc.edu or mkarwowski@valenciacc.edu

Thank you for your time and input!